16.95

HOMES APART

Anthony Lemon is a lecturer in Geography at Oxford University and a Fellow of Mansfield College. He is a southern African specialist, and has published widely on the urban, social and political geography of apartheid. He is author of *Apartheid: a Geography of Separation* (Saxon House, 1976) and *Apartheid in Transition* (Gower, 1987), and co-editor of *Studies in Overseas Settlement and Population* (Longman, 1980).

HOMES APART
SOUTH AFRICA'S SEGREGATED CITIES

Edited by
ANTHONY LEMON

P·C·P
Paul Chapman
Publishing Ltd

INDIANA UNIVERSITY PRESS
Bloomington and Indianapolis

DAVID PHILIP
Cape Town

This edition first jointly published
in the UK by
Paul Chapman Publishing Ltd
144 Liverpool Road
London N1 1LA
UK

in North America by
Indiana University Press
10th and Morton Streets
Bloomington
Indiana 47405
USA

and in southern Africa by
David Philip Publishers
208 Werdmuller Centre
Claremont 7700
South Africa

British Library Cataloguing in Publication Data

Homes apart: South Africa's segregated cities.
    1. South Africa. Coloured persons. Racial discrimination
    I. Lemon, Anthony
    305.8968

Library of Congress Cataloging-in-Publication Data
Cataloging information for this book is available from the Library of Congress
(91-72181).

ISBN 1—85396—117—5 Paul Chapman Publishing
ISBN 0—253—33321—0 Indiana University Press
ISBN 0—86486—199—0 David Philip Publishers

Typeset by Setrite Typesetters, Hong Kong
Printed and bound by Athenaeum Press Ltd, Newcastle upon Tyne

A B C D E F G H   8 7 6 5 4 3 2 1

4.9.92

CONTENTS

LIST OF CONTRIBUTORS

Anthony J. Christopher is Professor and Head of the Department of Geography at the University of Port Elizabeth. He has published six books and numerous articles in international journals on the historical and political geography of southern Africa, Africa and the British Commonwealth. He has recently made a special study of levels of urban segregation in South Africa.

Gillian P. Cook gained her doctorate at the University of Chicago and has lectured at Rhodes University, Grahamstown, South Africa and Cape Town. Her publications are mainly concerned with South African urban and settlement geography, and she is also interested in gender issues in geography. She is currently a Visiting Fellow at Southampton University.

Ron Davies is Professor of Geography, University of Cape Town, and was formerly Professor of Geography at the University of Natal, Durban. He has long been a student of the South African city, and in particular of the Durban metropolitan area.

Neil Dewar is a senior lecturer in Geography at the University of Cape Town. He holds Masters degrees in both geography and urban and regional planning, and was awarded his doctorate for a thesis on the changing geography of public finance from Salisbury to Harare.

James Drummond studied at the University of Glasgow and obtained his MA degree in 1984. Since then he has lectured at the University of Bophuthatswana in Mmabatho, South Africa, whilst pursuing part-time postgraduate research at the University of the Witwatersrand, Johannesburg.

Roddy Fox is a graduate of Strathclyde University, and spent several years lecturing at the University of Kenya in Nairobi before moving to his present lectureship in the Department of Geography at Rhodes University, Grahamstown, South Africa. He has published mainly on the urban and settlement geography of East Africa, but has now extended his research to South Africa.

Phillip S. Hattingh is Professor of Geography and Head of Department at the University of Pretoria. His research, spanning more than two decades, has mainly concerned black South Africans: the 'homelands', population and urbanization issues, recreation and also the spatial implications of public decision-making, policy formulation and execution.

Andre C. Horn is a lecturer in Geography at the University of Pretoria. His interests include urban geography, behavioural and humanistic geography. His current research focuses on the structural and human ecology of South African cities and its impact on human well-being.

D. S. (Skip) Krige is a senior lecturer in Political Geography at the University of the Orange Free State in Bloemfontein. He studied at the Universities of Stellenbosch, Oxford and the Orange Free State. His research interests are in political geography, the impact of apartheid planning, housing and informal settlements, and the transformation of South African cities.

Etienne Nel is a lecturer in Geography at Rhodes University, Grahamstown, South Africa, where he also graduated. He has previously been employed by the Natal Education Department, the University of Transkei and the Giyani College of Education in Gazankulu. He has recently completed a Masters thesis on racial residential segregation in East London.

Susan Parnell is a lecturer in the Department of Geography at the University of the Witwatersrand. She has published numerous papers on the historical and contemporary geography of South African cities, especially Johannesburg.

Gordon Pirie is a lecturer in Human Geography at the University of the Witwatersrand, Johannesburg. He has published extensively on the historical, urban and transportation geography of South Africa, and was awarded his doctorate for an historical study of racial discrimination in South African passenger transport.

Claudia M. Reintges is a lecturer in Geography at Rhodes University, Grahamstown, South Africa. Her research interests are mainly concerned with South African urbanization, especially the future of peri-urban settlements and forms of the local state in urban and rural areas.

David Simon lectures in Development Studies and Geography at Royal Holloway and Bedford New College, University of London. He is co-author of *The British Transport Industry and the European Community* (Gower, 1987), editor of *Third World Regional Development: a Reappraisal* (Paul Chapman, 1990), and has published extensively on transport, urbanization, planning and development issues. His Third World research interests have focused primarily on southern Africa, and especially Namibia.

Trevor M. Wills is a senior lecturer in Geography at the University of Natal, Pietermaritzburg, his home city in which he has developed a close interest. He is an urban geographer whose research has included involvement in planning projects in both the city of Pietermaritzburg and the province of Natal.

PREFACE

Already some three-fifths of all South Africans live in cities, and the proportion who do so will undoubtedly continue to increase in the post-apartheid era. The new urbanities will be overwhelmingly black, but they will come to cities which have been designed by, and primarily for, white people whose dominance in the social formation has been reflected in the colonial, segregation, and apartheid cities which they have successively created. Of these, the apartheid city was, like the wider society to which it belongs, the most deliberate creation. The extraordinary recasting which it represents reflects the Afrikaner Nationalist urge 'to be the architect and engineer of its own fundamentals' (de Klerk, 1975, p. 339).

As the end of the apartheid era approaches, South African cities are already undergoing rapid changes in response to multiple and growing pressures. Some of these pressures have already been the subject of legislative and policy changes, and others will be soon. In most cases such changes constitute a belated adjustment to what is already happening on the ground, often combined with new attempts to control and guide it.

Before very long, a new government representing the black majority will be faced with the formidable problems of South Africa's cities, combining as they do the planning problems of both First World and Third World urbanism and the huge headaches of an apartheid inheritance. Will the new authorities seek to restructure the cities as fundamentally as their predecessors? The enormity of such a task, after four decades of urban growth, would alone render such a task impossible. But changes there will clearly be, as the attempt is made to erase the human miseries of the apartheid city and to plan for the welfare of all urban dwellers instead of a dominant minority.

This is, then, an important time both to place on record the cities created by segregation and apartheid, and to take stock of those cities as they enter a new era of change. Much in their experience is common to all, as recent or forthcoming works on South African urbanization make clear (Tomlinson, 1990; Smith, 1991). Yet in a book concerned primarily with the design and form of cities as places to live, it is important to recognize what is distinctive as well as what is shared. The attempt to impose a common model on all cities has not produced uniformity on the ground: physical, historical, economic, social and political factors have ensured continuing differentiation, posing different problems which will require distinctive approaches to change. Conversely, where experiences are shared, the lessons learnt from policies applied in particular cities may be of benefit to those with comparable problems.

This volume brings together the experiences of 12 southern African cities, through the eyes of practising geographers whose common discipline provides unity (but not uniformity) of approach. Most have lived in and/or researched on the cities about which they write for long periods. It is sad, and of course significant, that all the geographers most qualified to write about the cities in question are themselves white; to date South Africa has produced almost no black urban geographers (and only a handful of black geographers overall) despite the preponderance of blacks in the urban population. If this book has a sequel a decade from now, its authorship will surely be more representative of those who dwell in South Africa's cities.

Nine of South Africa's major cities are represented here, including the three which share the functions of a capital city (Cape Town housing the legislature, Pretoria the executive and Bloemfontein the judiciary), the financial capital and largest city (Johannesburg), the four provincial capitals, the major port cities, and the first mining city, Kimberley. Three other southern African cities, which have proceeded further along the road to desegregation, are also considered for the relevance of their experience to post-apartheid cities. One of them, Mafikeng-Mmabatho, is South African in all but name, but its status as the capital of a nominally 'independent' state, Bophuthatswana, has permitted legal desegregation. Windhoek is capital of the newly independent state of Namibia, but during the period under examination here it was under South African control, and the effects of a partial dismantling of urban apartheid were carefully watched in Pretoria. Harare, which as Salisbury experienced settler-colonial policies not unlike urban apartheid, has completed a decade as capital city of a fully independent Zimbabwe.

In dealing with the present and looking to the future, all contributors faced common problems. Available data are frequently limited or unsatisfactory: thus township populations are invariably understated in census figures, and estimates of informal and squatter populations vary greatly. The pace of change since the September 1989 election has been so rapid that academic analyses of the present and prognoses for the future are in constant danger of being overtaken by events. The future will be influenced by many imponderables, and detailed prognostication can only be speculative. At the least, however, this volume seeks to provide the post-apartheid planners with a record of what they will inherit and how it came to be.

The book was conceived (and christened) while returning to Johannesburg with Gordon Pirie, after a wet weekend spent walking in the Drakensberg. Happily Gordon's enthusiasm, without which the idea would have progressed no further, was shared by others with whom I discussed the project, many of whom subsequently became contributors. Gillian Cook deserves special thanks for finding the time to comment very helpfully on my own chapters when she was in the process of settling in a new home in Britain. I am grateful to all the contributors for their co-operation and encouragement, and hope that the outcome does justice to the intentions we shared.

Anthony Lemon
Oxford
1 October 1990

NOTES

Ethnic terminology

Frequent reference to the ethnic groups officially classified under the Population Registration Act of 1950 is unavoidable, but this does not reflect acceptance of such classification. The term 'black' is used to include those classified as coloureds, Asians as well as blacks, as all these groups regard themselves as black. 'Indian' is preferred to 'Asian', except when reference is made to Chinese as well as Indians. To avoid confusion, those officially classified as black in South Africa are referred to here as 'Africans', but in Chapter 13 the normal Zimbabwean terminology, 'blacks', is used.

Currency

The value of the Rand has fluctuated considerably since its creation at R2 = £1 in 1961. From R1.50 = £1 in 1983 it has declined, precipitously in 1985−6, to almost R5 = £1 in 1990.

1
THE APARTHEID CITY
Anthony Lemon

From colonial to apartheid cities

In the absence of an indigenous urban tradition, South African cities were established by white settlers, who regarded the cities as their cultural domain. From the outset they needed black labour, but were reluctant to accept those who provided it as fellow citizens. The attempt 'to secure labour-power without labourers' has thus been a major feature of state urban policy (Maylam, 1990), leading to both controls on rural–urban migration and segregation of those whose presence in the 'white' cities was unavoidable.

Three main phases may be identified in the development of urban policy and practice in South Africa (Davies, 1981). The 'settler–colonial' period lasted from the beginnings of white settlement at the Cape in 1652 until the early years of Union after 1910. The Natives (Urban Areas) Act of 1923 marked the beginning of the conscious nationwide pursuit of urban segregation, which was to be superseded by the more rigid and far-reaching policy of urban apartheid from 1950 onwards, symbolized above all by the Group Areas Act of that year. These legislative landmarks provide convenient boundaries, but there are strong threads of continuity between the three phases – so strong, indeed, that they are likely to exert a profound influence even on the post-apartheid city which has begun to emerge since 1979.

The present work focuses primarily on the apartheid city, a term which suggests a phenomenon unique to South Africa. This would be only a half-truth. No other country, certainly, has embarked on so thorough a reorganization of its urban space for the purposes of segregation. But the apartheid city cannot be cut off from its roots, which Christopher (1983) even traces back to thirteenth-century English colonial practice. Until at least the 1950s, South African cities could easily be viewed in the context of African and colonial cities generally (Simon, 1984a). Particular parallels between South African cities and Nairobi have been noted (Fair and Davies, 1976; Western, 1984), while Winters (1982) describes French and Belgian colonial planners' techniques of residential zoning which closely resembled apartheid town planning. Abu-Lughod (1980) even subtitles her study of Rabat 'Urban Apartheid in Morocco', and Western (1985) extends the comparison to one between Cape Town and the ex-colonial port cities of China.

Certain key features are common to the social formation of all colonial and settler-colonial cities, and in South Africa to the post-colonial cities of the segregation and apartheid phases. Forces of control, imposed to maintain relations of dominance, depended crucially on control of access to political power, but also included control of access to means of production and levels of employment, and to the means (education, training and opportunity) of upward socio-economic mobility; control over land resources, their ownership, use and distribution, and over access to services and amenities; and control over spatial relations through segregation and urban containment. These controls were used, *inter alia*, to achieve formal urban development, to contain forces threatening to the economic and social interests (including health) of the dominant group, and to ensure that subordinate group housing was self-financing. Such controls, all characteristic of both segregation and apartheid cities in South Africa, were underpinned by appropriate administrative and policing mechanisms. Crises that arose as outcomes of incomplete control were met by progressive intensification of the means of control designed to maintain the social order, coupled with minor concessions.

Societies held together by minority dominance tend ultimately to reach a conflict threshold at which structural changes become necessary. Recognition that South Africa is fast reaching this point distinguishes the Presidency of F. W. de Klerk from that of his predecessor, P. W. Botha. The latter's reforms were imposed rather than negotiated, and characteristically took the form of adaptations of the existing order in order to lessen conflict and increase legitimacy, but without altering the fundamental dominant-subordinate relations of the social formation. As in other colonial societies, such measures inevitably proved ineffective, and must eventually be succeeded by fundamental political change. The structural changes which follow will depend on the political and economic order adopted and, crucially, the degree to which inherited structures are amenable to transformation. It is to those structures which we now turn.

Early urbanization and the beginnings of control

African urbanization proceeded slowly in the second half of the nineteeth century, and even in the first two decades of the twentieth century the percentage of urbanized Africans remained fairly constant at 12 to 13 per cent (Shannon, 1937). The total urban African population rose from 336,800 in 1914 to 587,200 in 1921, with a high proportion of males reflecting the importance of migrant labour, especially on the mines. Manufacturing had not yet developed significantly, and most Africans still relied on the rural economy for subsistence. The white urban population of 847,400 comfortably exceeded the African figure in

1921, but whites were already 55 per cent urbanized.

The earliest urban locations were established by missionary societies as a means of protecting their African and coloured congregations from the excesses of internecine conflict and exploitation (Christopher, 1989a). No centralized state control was exercised over African urbanization before 1923, and even municipal controls were absent through most of the nineteeth century. A high degree of municipal autonomy allowed the development of a variety of approaches towards segregating and controlling Africans, but the major cities failed to develop any comprehensive policies for housing the urban African population (Maylam, 1990). In the mines, however, the compound system introduced in Kimberley in the 1880s rapidly became the model for mines and mining towns throughout southern Africa (Mabin, 1986). The compounds assured the mine-owners of the supply of experienced labour they needed, but prevented the emergence of an organized working class (Turrell, 1984). The system was adopted for other workers too, including those employed by municipalities, factories, warehouses and docks.

Even in the mining towns, however, large numbers of Africans lived outside the compounds. In the Johannesburg of the 1920s, some 60,000 Africans did so: 5,500 were housed in municipal townships, 6,000 in freehold or leasehold townships on privately owned land, and others lived as tenants on white-owned properties, occupying shanties, out-rooms or disused warehouses or workshops let by slum landlords (Maylam, 1990, p. 60).

Such African settlement was largely uncontrolled; segregationist pressures grew only slowly in most cities, and segregated townships remained few in number. In Durban the mayoral minutes of 1887 report a public meeting which discussed assaults and other crimes by Africans and asked that locations should be established at a convenient distance from the town (Kuper, Watts and Davies, 1958, p. 30). Unlike Cape Town, however, Durban built no major locations before 1923, although it did require casual, or 'togt', workers to live in compounds from 1903 (Maylam, 1990).

Durban Indians, who were seen as a more direct threat to whites, attracted a more vigorous municipal response based on residential segregation, municipal exclusion, and commercial suppression (Swanson, 1983). Indians were restricted to separate 'bazaars' in many Transvaal towns in the 1890s, and restrictions were tightened thereafter. The Orange Free State pursued the most deliberately segregationist approach. It excluded Indians altogether from 1891 (a law which was repealed only in 1985), and decreed in 1893 that only whites could own or lease fixed property in Free State towns. Only in the Cape was relatively free movement permitted and greater integration evident, especially between whites and coloureds.

In the private housing market, segregation was commonly achieved through the inclusion of racial exclusion clauses in suburban property deeds. This device was widely adopted by developers on the Rand in the late nineteenth century, and emulated elsewhere, especially after the First World War. However, enforcement of such private conditions of title was subject to the civil law and not the criminal code, which often made them difficult for either developers or local residents to enforce (Christopher, 1989a).

White fears of the spread of contagious disease were a significant impetus to municipally ordained segregation. The establishment of locations in Cape Town, Port Elizabeth and Johannesburg was precipitated by a frantic effort to limit bubonic plague (Davenport, 1971). Unfortunately such concerns did not seem to guide the siting of the locations, many of which were close to sewage farms and refuse dumps. Not surprisingly the locations themselves soon became a problem, and in 1914 the Tuberculosis Commission reported that the conditions in which tuberculosis flourished were widespread in locations throughout the country. A bad influenza epidemic in 1918 revealed the distressing conditions in which Africans lived and the health threat which they posed. In Johannesburg this aroused the civic conscience and led to the first African housing schemes, but it is indicative of contemporary attitudes that 'native locations' were the responsibility of the city's parks department in the 1920s (Lewis, 1966).

Segregated cities, 1923–50

During these three decades, the combination of deteriorating economies in the African reserves and the growth of manufacturing employment in the cities led to dramatic growth in the number of urban Africans, from 587,200 in 1921 to 1,146,700 in 1936 and 2,329,000 in 1951. The percentage of urbanized Africans doubled to 28 per cent (Maylam, 1990), and the proportion of Africans living in family as opposed to migrant conditions steadily increased. The response of the authorities to the housing needs generated can be judged from the fact that only 99,250 houses were built between 1920 and 1950 for all race groups, half of them for Africans (South Africa, 1987a, p. 58).

State intervention to deal with these changing circumstances began in 1923 (Morris, 1981b). Legislation to regulate the conditions of urban Africans was envisaged as early as 1912, and a draft Urban Areas Bill embraced the conception, championed by the mining industry, of an African middle class with property rights in urban areas. Instead, the ideas of the Stallard (Transvaal Local Government) Commission of 1922, concerned to restrict the number of urban Africans so as to minimize expenditure on their locations, won the day, with the assistance of both white labour and urban commercial capital which saw their position as threatened by a growing African influx (Rich, 1978). The essence of 'Stallardism', which was to form the basis of official attitudes for several decades, is summed up in the oft-quoted dictum that 'the native should only be allowed to enter the urban areas, which are essentially the White man's creation, when he is willing to enter and minister to the needs of the White man, and should depart therefrom when he ceases so to minister' (Transvaal, 1922, para. 42). This doctrine has had far-reaching implications for the provision of services, property ownership, participation in administration, and the morphology of African townships (Smit and Booysen, 1977).

The 1923 Natives (Urban Areas) Act embodied Stallardist principles. It empowered, but did not compel, local authorities to set aside land for African occupation in segregated locations, to house Africans living in the town or require their employers to do so, and to implement a rudimentary system of influx control. Municipalities were required to keep native revenue accounts, and the revenue accruing from fines, rents and beer hall profits had to be spent on the welfare of the location. The Act also provided for an embryonic form of consultation through advisory boards, and sought to control location brewing and trading.

Some municipalities, especially the larger ones such as Johannesburg, Kimberley and Bloemfontein, adopted the Act almost immediately, although Durban did so only in the early 1930s. Smaller towns, fearing excessive financial responsibilities, took longer, but most locations had been registered by 1937. Few local authorities, however, were prepared to subsidize African housing from general revenue in the inter-war years, and the income from native revenue accounts, though often substantial, never proved sufficient to meet the housing needs of Africans. Many municipalities failed to establish advisory boards, and by 1937 only 11 towns had systematically instituted influx controls (Savage, 1986).

The importance of the 1923 Act rests not only in its pioneering nature, but in the framework it established for future legislation (Maylam, 1990). This embodied the central principles of segregation (and hence relocation) and influx control, together with the expectation that African urbanization would be self-financing and the co-optation of Africans, initially in an advisory capacity, to help the system to operate. All these principles were developed and strengthened in subsequent legislation, up to and in some respects including the Botha reforms of the 1980s.

Machinery providing for systematic influx control was introduced in the 1937 Native Laws Amendment Act. Under this measure, the implementation of which was still discretionary, Africans coming to the towns were allowed 14 days to find work (reduced to three days in 1945), and individual Africans might be 'rusticated' if municipal returns showed a labour surplus. Controls over the entry of women to urban areas were reinforced, but the government did not make women carry passes, without which the legislation proved difficult to enforce. The Act represented an increase in centralization, by allowing the Minister of Native Affairs to compel a local authority to implement any section of the 1923 Act, or to have the section implemented by his own department. Municipal autonomy was further weakened by the 1944 Housing Amendment Act which set up a National Housing and Planning Commission, with powers to intervene in local housing policy.

Hitherto, state urban policies had been concerned largely with Africans, but in the late 1930s whites became increasingly concerned as Indians began to 'infiltrate' predominantly white residential areas, especially in northern Durban (Chapter 12), and trading areas in Transvaal towns. The white reaction was disproportionate to the scale and nature of such movement, but led to two 'penetration' commissions and the imposition of new restrictions on Indian occupation and ownership of property in 1943 and 1946. These characteristically *ad hoc* measures on the part of Smuts' United Party government may be seen as a forerunner to the all-embracing changes to be ushered in by its National Party successor in 1950.

The 1940s were years of rapid change. Wartime shipping shortages deprived South Africa of many imports, and encouraged rapid industrial growth, which in turn increased labour demands. As industrialists began to stress the value of semi-skilled Africans in manufacturing employment and several official voices urged the need for reform, the government appeared slowly to be recognizing the strength of their arguments (Lemon, 1987). Smuts himself seems to have viewed African urbanization realistically, observing that 'You might as well try to sweep the ocean back with a broom' (1942, p. 10). Yet his government sought to strengthen influx control through the 1945 Natives (Urban Areas) Consolidation Act, Section 10 of which allowed an African to claim permanent residence in an urban area only if he had resided there continuously since birth, had lawfully resided there for 15 years, or had worked there for the same employer for 10 years. This draconian law was, however, only applicable at the request of a local authority until the Nationalist government made it mandatory in 1952.

During the Second World War, manpower and materials were concentrated on wartime needs and housing lagged badly behind. The situation worsened during the early post-war years of continued rapid industrial growth, and squatter settlements mushroomed around South African cities. Most made some attempt to catch up on the housing backlog, but several factors combined to prevent major efforts in this direction, including the sheer immensity of the task, reluctance to increase the rates to subsidize large-scale African housing, and continuing doubts about the permanency of the urban African population, encouraged by uncertainty over the life of some of the goldfields.

Meanwhile growth of the squatter settlements produced increasing bitterness and a more radical calibre of black leaders. The Fagan Commission (South Africa, 1948a) strongly criticized the migrant labour system and the inhumanity of the pass laws, but the dramatic growth of urban African numbers alarmed whites and became a central issue in the 1948 election campaign, in which the National Party's use of *swart gevaar* (black danger) arguments contributed to its unexpected victory. Segregation was soon to give way to apartheid, as Afrikaners sought to manifest their dominance yet more strongly in the fabric of the cities. Had the United Party won the 1948 election, both urbanization and urban residential patterns would no doubt have been treated more pragmatically; instead it has taken 40 years for the realities of social and economic change to challenge the artificiality of apartheid cities.

Davies' (1981) model of the 'segregation city' is a useful representation of South African cities before they were subjected to apartheid planning (Figure 1.1). It incorporates a central business district (CBD) which includes a small Indian CBD on the edge of the white business area. This reflects the fact that, where they were sufficiently numerous, Indians established a significant presence in many CBDs, most notably in Durban (Grey Street), Pietermaritzburg, and Johannesburg (Pageview). Coloured people never managed to do this, and

THE SEGREGATION CITY

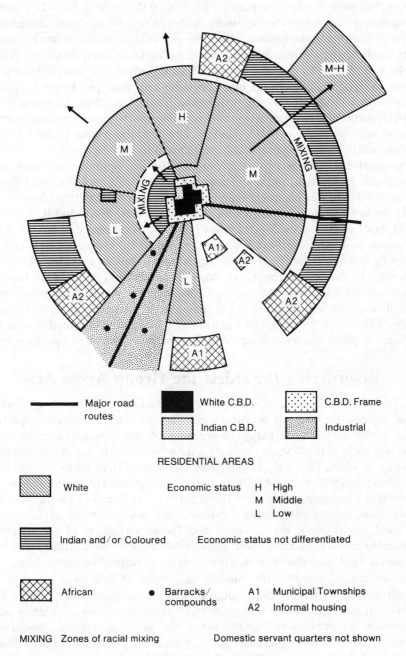

Figure 1.1 The segregation city (after R. J. Davies (1981))

Africans were denied the opportunity. Whites occupy most of the residential space in Davies' model, with a broadly sectoral differentiation of income groups, but with shortage of space having forced the most recent white housing developments to move out beyond a hitherto peripheral band of Indian or coloured housing. Low-income white housing is close to the industrial sector, as is much of the African housing, while barracks or compounds for African workers are located within the industrial zone. Indian and coloured housing is mainly peripheral, but also includes an inner city zone just beyond the CBD frame, and 'islands' within otherwise white residential areas. Importantly, the model incorporates two extensive zones of mixing between whites and coloureds; central suburbs such as South End in Port Elizabeth and District Six and Woodstock in Cape Town would have fallen into this category.

Pre-apartheid cities were, as the model suggests, highly but not completely segregated. In Durban, indices of segregation for 1951 included the following (where 1.00 indicates complete segregation): Indian/white 0.91; Indian/African 0.81; coloured/white 0.84; African/white 0.81 (Kuper, Watts and R. J. Davies, 1958). For the same year, W. J. Davies (1971, p. 148) calculated indices of 0.89 for whites and others in Port Elizabeth, and 0.80 for coloureds and Africans. These figures are higher than those calculated by Duncan and Davis (1953) for American cities, despite the use of smaller enumeration areas in the latter. In the South African cities African/white segregation would have been almost total if domestic servants had been excluded. What need, then, to segregate further? The extensive re-zoning and wholesale movement ordained by apartheid planners implies that something more than mere segregation was intended.

Re-ordering the cities: the Group Areas Acts

The Group Areas Acts of 1950 and 1966 have had more far-reaching effects on racial segregation than any previous legislation, producing distinctive apartheid cities which represent a major re-ordering of the segregation cities which preceded them. For 40 years these Acts have been a cornerstone of apartheid, and one to which Mr P. W. Botha clung tenaciously right up to the end of his Presidency in 1989. It is not difficult to see why: group areas form the basis of segregated education, health and social services, and also of local authorities. In addition, race zoning assumed a new importance with the introduction of the 1983 constitution; the coloured and Indian chambers of the tricameral parliament cannot exercise their responsibilities for 'own affairs' without a territorial base, and that base is effectively their respective group areas.

Group areas exemplify the fundamental tenet of apartheid ideology that incompatibility between ethnic groups is such that contact between them leads to friction, and harmonious relations can be secured only by minimizing points of contact. By preventing contact, urban residential segregation hinders any transition from conflict pluralism to a more open pluralistic society. Such a transition depends upon the development of cross-cutting cleavages – individual allegiances and affiliations which cut across ethnic divides – and such links are most likely to be established within the neighbourhood or local community.

Race zoning greatly inhibits even the limited inter-group social contact which might naturally occur (given a degree of black socio-economic mobility) in churches, sports clubs and especially schools. Multiracial gatherings have needed the organization of transport for the blacks and, frequently, application for special permits. In apartheid's 'world of strangers', moreover, such social gatherings have tended to be self-conscious, and it is almost impossible to be truly 'colour-blind' in human relationships. As the novelist Nadine Gordimer puts it,

> In the few houses in Johannesburg where people of different colours met, you were likely to meet the same people time after time. Many of them had little in common but their indifference to the different colours of their skins; there was not room to seek your own kind in no man's land: the space of a few rooms between the black encampment and the white.
>
> (Gordimer, 1962, p. 168)

For the majority, then, race zoning has kept people from knowing or understanding one another. Blacks, especially Africans, have always entered white homes as servants, but many whites have never visited an African township, and remain comfortably ignorant of the conditions in which the majority of people live.

Group areas legislation provided for the extension throughout South Africa, and to all race groups, of the land apportionment principle long existing in the African reserves, and since 1923 (depending at first on local authority compliance) for urban Africans. The operation of group areas legislation was essentially urban, as many of its provisions applied elsewhere already. The 1950 Act imposed control of inter-racial property transactions and inter-racial changes in occupation of property, which were subject to permit. No less than 10 kinds of area were defined, but the ultimate goal was the establishment of areas for the exclusive occupation of each group. Group Areas might be proclaimed in respect of either ownership (with controlled occupation) or occupation (with controlled ownership), but the proclamation applied to both ownership and occupation in the final form. 'Border strips' might be designated to act as barriers between different group areas to ensure that no undesirable contiguity occurred. 'Frozen zones' might be proclaimed if an area was considered suitable for proclamation but not immediately required: control would then be exercised over the use and development of the area, including the freezing of property transactions, to facilitate its eventual proclamation as a group area. 'Future border strips' might likewise be set aside with future needs in mind. Control over occupation could be temporarily withdrawn at any time by special proclamation; the possibility of establishing such 'open' areas, as they came to be known, caused much controversy between several local authorities and the Group Areas Board, especially when it was decided to retain a black (usually Indian) commercial district in what had been proclaimed a white group area (Lemon, 1987).

Implementation was the responsibility of the Group Areas Board and, subsequently, the Community Development Board. The former, in order to provide effective recommendations, needed the assistance of experienced surveyors, engineers and planners, which demanded the co-operation of local authorities. Where this was refused, as in Cape Town, local authorities risked the imposition of an arbitrary zoning plan. When, after opportunity for objection and inquiry, the government finally approved any recommendations, their implementation was a matter for the Community Development Board, which dealt with housing, the development of group areas, the resettlement of dislocated persons, slum clearance and urban renewal.

Group areas legislation radically extended control over private property. The Group Areas Development Act of 1955 provided machinery for compensation, but established procedures for regulating the sale price of property in the open market, and provided for expropriation of properties under a system of public acquisition for group area development. White acceptance of such measures clearly rested on the assumption that others would be the victims. That this has proved to be the case is hardly surprising since whites passed the legislation, whites implemented it, whites alone were represented on local councils, and whites were generally far better able to defend their interests than other groups. In most cases the inner city and suburbs were proclaimed white and other groups consigned to the periphery, and no less than 99.7 per cent of whites already lived in what became white group areas (Christopher, 1989a). The vast majority of the 125,000 *families* moved in terms of group areas planning have been coloureds and Indians. In addition, many Africans have been indirect victims of group areas planning; they are excluded from official statistics of group area removals because they were moved in terms of amendments to existing legislation, without even the semblance of consultation embodied in the Group Areas Acts. Platzky and Walker (1985) suggest that 730,000 Africans were resettled in urban areas between 1960 and 1983, and the total figure for 1950–90 almost certainly exceeds one million.

Ironically, the implementation of group areas legislation led to the allocation of far greater financial resources to black housing than hitherto. They were directed primarily to the fulfilment of an ideological commitment rather than to housing problems *per se*, thus consuming resources which could have been used to reduce the growing housing backlog. In some cases people were physically better housed after removal, at least until their often jerry built houses deteriorated, but in other cases race zoning has exacerbated overcrowding and squalor.

Housing conditions, removals and segregation can all be quantified, but the human damage inflicted by race zoning is immeasurable. It has unquestionably accentuated and even initiated the racial antipathy it was (supposedly) intended to prevent (Pirie, 1984c). Irrespective of physical improvement or deterioration of housing conditions, many communities were emotionally impoverished by the destruction of their community and their remoteness from the area in which they had grown up. Thus Hart and Pirie (1984) describe the paradox of 'emotional plenty among the material shortage' of Sophiatown, Johannesburg, before its African inhabitants were removed. But it is appropriately Cape Town, the city most affected by group area removals, which is the subject of the most penetrating study of their devastating human impact (Western, 1981). Probing the images of the city held by its coloured people, Western demonstrates the humiliating damage which removal and destruction of homes had on self-esteem and security, and the way in which people had to learn again their place on the edges of a callous urban society. By engaging powerfully with people and place, Western gives flesh to his conceptual theme, which remains *the* theoretical message of apartheid cities: 'human social relations may be *both* space forming *and* space contingent' (p. 5).

The space formed has also been modelled by Davies (1981), and is illustrated in Figure 1.2, which may be compared with the segregation model already described (Figure 1.1). An exclusively white CBD is surrounded by an extensive consolidated white residential core with freedom to expand in environmentally desirable and accessible sectors in suburban localities. Socio-economic patterns within white residential areas remain relatively undisturbed. Coloured and Indian group areas, and especially African townships, are located peripherally within given sectors; hostels for migrant workers no longer adjoin the workplace but have been relocated within these townships. Pushing the poor to the periphery has invariably much increased their journeys to work, especially where African townships have been constructed across bantustan borders.

How far was this model translated into reality? The 1950 Group Areas Act was only made effective after its amendment in 1957, so its real impact was only felt in the 1960s and 1970s. This impact was not uniformly drastic, and the imprint of the colonial heritage has survived 80 years of segregation and apartheid in some places (Christopher, 1990). In many towns small communities of coloureds and Indians were not immediately affected as they were deemed too insignificant to warrant a separate group area. Sometimes group areas were deproclaimed where the community was small, and it was resettled in a larger town (Christopher, 1991). On the eastern Witwatersrand in the 1960s attempts were made to concentrate coloureds at Boksburg and Indians at Benoni but the proclamation in the 1980s of some 57 new coloured and 15 new Indian group areas suggested that such earlier policies of concentration had not succeeded (ibid.). Even today implementation is not complete, and Christopher (1989a) notes nine group areas intended to house small Indian communities which still await the construction of housing.

By the end of 1987, however, more than 1,300 group areas had been proclaimed. Whites, who numbered 67 per cent of the non-African urban population in 1985, had been allocated 87 per cent of the total land in group areas. Christopher (1989a), in a revealing analysis of enumeration district statistics for the 1985 census, found that some 10 per cent of the urban population (including Africans) still lived outside its designated group area (Table 1.1). In contrast to earlier periods, the Western Cape emerged as the most segregated region, with only 5.7 per cent of its urban population living outside its designated area. This is partially due to the removal of coloureds from African townships (also common in the Eastern Cape), whereas in other regions, notably the Orange Free State, this was regarded as unimportant. As a result, coloured-African indices of dissimilarity were considerably higher in the

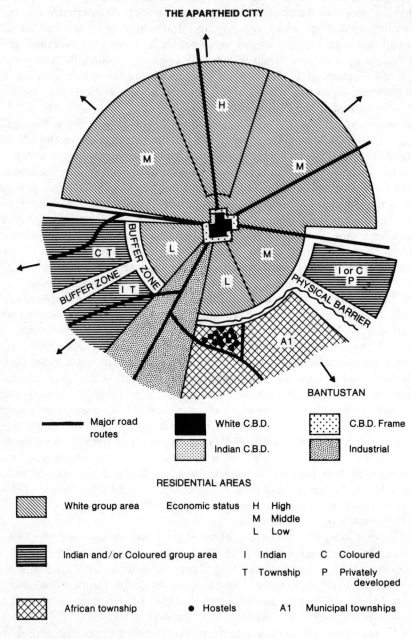

THE APARTHEID CITY

Figure 1.2 The apartheid city (after R. J. Davies (1981))

Table 1.1   Urban population of South Africa, 1985: group/development areas of residence

Population group	Area of residence				Total
	White	Coloured	Indian	African	
White	4,078,634	6,353	2,339	4,741	4,092,067
Coloured	189,033	1,978,257	17,263	48,038	2,232,591
Indian	44,818	10,839	710,208	1,680	767,545
African	820,208	24,140	25,552	4,028,390	4,898,270
Total	5,132,693	2,019,589	755,362	4,082,849	11,990,493

(*Source*: Christopher, 1989a)

Cape than elsewhere (ibid.). The Western Cape also reflects the removal of coloureds from once multi-racial suburbs designated white, such as Cape Town's District Six (Chapter 2). The dissimilarity index between whites and coloureds in Cape Town as a whole rose from 51.52 in 1921 to 95.13 in 1985, by which time it was the most segregated major city, with only 5.0 per cent of its people living in the 'wrong' group area (ibid.).

Segregation indices between whites and other groups appeared deceptively low in 1985, dropping below 0.75 in 10 per cent of towns and even below 0.5 in a few towns (Christopher, 1989a). In some cases, mainly in Natal and the Transvaal, these figures are misleading, as they exclude most of a town's African population because it lived in nearby bantustans. Otherwise, the main reason for low segregation indices is the housing of African workers in white areas, mainly live-in servants but also including migrant workers in mining compounds, who made up a quarter of the total. State institutions such as hospitals, prisons, military and police installations are also found mainly in white areas, but frequently cater for the entire population and thus produce a measure of statistical integration which, as Christopher (1989a, p. 259) drily observes, 'may be belied by their actual management'.

Indian trading, CBDs and group areas

Whereas all groups except white suffered residential disruption and community dislocation as a result of race zoning, the economic effects fell disproportionately on Indians owing to the extent of their commercial involvement (Lemon, 1987). Afrikaners have tended to regard Indians as parasitical due to their association with commerce, and Fatima Meer probably voiced the feelings of many in her community when she asserted that 'one of the prime purposes of the Group Areas Act is to eliminate, or at least to reduce to a minimum, Indian commerce' (1971, p. 23). Dislocation has been particularly great in the Transvaal and the Cape, where absolute numbers of Indians are relatively small but the proportion dependent on commerce was very high. By 1966, only 7.5 per cent of Transvaal Indians remained unaffected by group area proclamations, and even in Pretoria, where nearly one-third of Indian traders were unaffected, the removal of African and coloured populations living near the Asiatic bazaar made loss of business inevitable. In Pageview, Johannesburg, where Indian traders had used every device including Supreme Court action to fight removal, the government eventually assisted the city authorities in building an Oriental Plaza, but the level of rents prevented many small traders from re-establishing their businesses.

Indian CBDs have survived only in the Natal cities of Newcastle, Pietermaritzburg and Durban. In Pietermaritzburg, many Indian traders were fortunate in that their CBD formed the apex of a wedge of the city centre proclaimed for Indian and coloured residential use. This was not the case in the country's largest Indian trading area, in and around Grey Street in Durban. After years of damaging uncertainty and planning constraints, Grey Street was proclaimed an Indian Group Area for trading and light industrial purposes in 1973, but its 13,000 residents were required to move; not all of them had actually done so when the decision was reversed a decade later. Indians formerly trading outside the three surviving CBDs, and those in many smaller Natal and Transvaal towns, lost their livelihood when they were removed to group areas. In small places, removal two or three kilometres from the centre of town killed businesses which depended largely on white and African custom (Maasdorp and Pillay, 1975).

A limited step towards non-racial trading areas was taken in the 1966 Group Areas Act, section 19 of which allowed the proclamation of group areas for a specific purpose or use, instead of for ownership or occupation of a specific group. Use of this provision was slow initially, but there were 26 such areas by 1983. Most were small in size, and three cities — Durban, Port Elizabeth and Kimberley — actually had two separate 'section 19' areas. The procedures were cumbersome, and it took Pietermaritzburg City Council five years to negotiate the establishment of a free trade area in upper Church Street, an area once owned and occupied by Indians, many of whom had continued to own properties and trade there by nominally registering under white ownership.

Ultimately the widespread use of this unofficial 'nominee' system helped to encourage legalization of open trading areas. In terms of the Group Areas Amendment Act of 1984 local authorities, other organized bodies or the Minister could submit requests to have areas investigated for the purpose of having them declared free trading areas. Despite the cumbersome procedure involved, 90 CBDs (but no suburban shopping centres) had been opened for trading by all races at the end of 1988, nearly half of them in the Cape. Indians were the main beneficiaries; few Africans possessed the experience and capital to establish business in the CBDs, while other factors, including a movement of consumers from CBDs to suburban centres and a fear of boycotts and stayaways, may also have deterred them.

Group areas under pressure

Legally enforced residential segregation came under increasing pressure in the 1980s. Whereas there was an estimated surplus of 37,000 housing units for whites in 1985 (a reflection of low natural increase and the beginnings of a white exodus), the shortage for coloureds was 52,000, for Indians 44,000, and for Africans officially 538,000, but in reality far more (de Vos, 1986a). With increasing numbers of middle-class blacks able to afford houses in white areas, this increasingly led to houses in those areas coming into black ownership, mostly via whites who acted, for a fee, as nominees on their behalf, or through closed corporation deals. In this way a new generation of 'grey areas', pockets of integrated residential settlement, was born in several major cities, especially Johannesburg (Chapter 9).

Such changes on the ground led to other pressures for the law to be changed. Pirie (1987) notes several examples of relatively liberal white local authorities such as East London, Durban and Pietermaritzburg not only accepting the emergence of grey areas *de facto* but seeking varying degrees of *de jure* change, and even threatening to defy the law. Such authorities tended to give unqualified support to the trickle of blacks who used their right to apply for government consent to live in the 'wrong' area, and some 90 per cent of the 2,716 applications made in 1985 were granted.

At the same time the strict application of the 1966 Group Areas Act was increasingly compromised by judicial decisions, most notably the Govender ruling by the Appeal Court in 1982 which gave discretionary jurisdiction to the courts on eviction orders, whilst it was also ruled that such orders could only be imposed when a full inquiry regarding all the circumstances in each individual case had first been undertaken. The stated considerations to be taken into account rendered convictions highly problematic, and must have influenced the announcement by the Attorney General of the Transvaal that he had of late not been instituting any prosecutions in terms of the Act (de Coning, Fick and Olivier, 1987).

A final challenge to official policy came from major business firms. In Johannesburg, particularly, some of them had begun to buy houses for senior black executives in the affluent northern suburbs, and at the Pretoria 'business summit' in 1986 the State President was faced with strong demands for the repeal or drastic amendment of the Group Areas Act.

In the face of these various pressures President Botha showed little flexibility, repeatedly emphasizing (no doubt with electoral considerations in mind) that an 'own community life' for the different population groups remained a corner-stone of National Party policy, and stressing the need for special protection for low-income whites in particular. This group, and especially its older members with limited mobility to move elsewhere, is certainly the most vulnerable to change. Like people in zones of transition in many parts of the world, they are compelled to cope with new conditions in the sphere of human relations, yet they are among the least well-equipped members of society to do this. Their vulnerability maximizes their social and political alienation in such circumstances.

None the less, the government must have realized its own interest in making it easier for middle-class and upwardly mobile blacks, especially Africans, to move out of segregated townships. As Sampson tellingly observes, 'the hope that the growing middle class would form a bourgeois buffer was negated by segregation which pushed graduates back among their own people' (1987, pp. 136−7). If only high-income suburbs were opened to blacks, the volome of migration and hence the 'bourgeois buffer', would be very limited.

The wheels of reform were set in 1981 when the Strydom Committee was appointed to advise the government on, *inter alia*, possible changes in group areas legislation. It recommended repeal of the 1966 Act and its replacement by a Land Affairs Act, in terms of which the ownership and occupation of land would be controlled by way of restriction in the title deeds (South Africa, 1983). The State President requested the President's Council to advise him on this report. Publication of the latter's report was delayed until after the May 1987 white election; meanwhile speculation focused increasingly on the idea of a 'local option', which appeared to answer many of the government's political fears. Low-income whites and others who felt most threatened could keep their neighbourhoods white, allowing the government to proclaim in conservative areas that nothing had changed, whilst in those cities which were already challenging the law the government would be able to maintain that it had responded to the claims of local democracy.

The local option concept proved central to the President's Council report (South Africa, 1987b), and was accepted by the government. Its legislative response was, however, characteristically equivocal. Three Bills were tabled, of which one, the Group Areas Amendment Bill, sought to strengthen enforcement of segregation by substantially increasing fines for contravention by landlords, vendors and residents, and by making it obligatory for courts to evict persons contravening the Group Areas Act, therefore overriding the Govender ruling. This Bill, rejected by the coloured and Indian houses of the tricameral parliament, was subsequently referred back for reconsideration by the President's Council, and dropped: a rare instance of the government not passing its intended legislation.

The other two Bills were passed. The Free Settlement Areas Act of 1988 enables the declaration of open areas on the recommendation of a Free Settlement Board. To change the status of an area an application has to be made to the Board by either the State President, the relevant Minister's Council, Provincial Administrator, or the local authority concerned; communities and individuals cannot apply directly to the Board. The remaining measure, the Local Government Affairs in Free Settlement Areas Act, provides for residents of free settlement areas (FSAs) to elect a non-racial management committee. However, those who were registered voters for a local authority have the option of remaining on its separate voters roll. This could bring about an anomalous situation in which a few conservative whites could effectively control an open area, as the local authority, unlike the area's management committee, has decision-making powers (Saff, 1990, Chapter 6).

The combination of the three Bills suggests that the Government's intention was to accept FSAs in neighbourhoods where integration had gone too far to reverse, and in some new developments, but to halt the process elsewhere before it was too late. In 1987 the police investigated 1,243 complaints in terms of the Act, but only three prosecutions took place (Urban Foundation, 1990a);

this rose to 1,689 complaints and 98 prosecutions in 1988 (ibid.) but it seemed clear that the Act was becoming a dead letter. In August 1989, however, with the following month's election clearly in mind, the government announced a new scheme encouraging whites to report black neighbours who were contravening the Group Areas Act to specially appointed officials in major centres. In practice, these officials have given blacks living in the 'wrong' area the correct forms for them to apply for permits to remain, and in mid-1990 the government was reportedly issuing permits by fax machine within 24 hours of application (*Star* International Airmail Weekly, 27 June 1990). In some instances officials even helped to protect blacks against both hostile white neighbours and exploitation by white owners charging exorbitant rents. Between July 1989 and February 1990 the police investigated 1,249 group areas contraventions, but not a single charge was laid (Urban Foundation, 1990a).

The establishment of FSAs was proceeding in 1990, but in an air of uncertainty after President de Klerk had announced his hope of repealing the Group Areas Act in 1991 (Chapter 14). The creation of these isolated areas of racially open settlement may lead to duplication of the overcrowded and deteriorating conditions which characterize the much publicized grey area of Hillbrow. The process whereby FSAs are declared could also encourage right-wing mobilization and worsen race relations. Perhaps influenced by such fears, the Nationalist members of the Johannesburg City Council joined forces with their Democratic Party (DP) colleagues in September 1990 to ask for the whole city to be declared a FSA. The DP-controlled councils of Sandton, Randburg and Midrand may well follow the same course once the consequences for local government structures are clarified. This approach seems more likely to win approval than the approach of the liberal Cape Town City Council, which rejected the Free Settlement Areas Act in 1989 and asked that the city be exempted from the Group Areas Act.

In the first new estates to be declared FSAs, such as Country View at Midrand, demand for properties was high. Whilst the opening of such areas allows buyers to move immediately, they are restricted in locational choice, and may find that prices drop when choice is no longer constrained by race zoning. The same may be true of blacks who have paid a premium to buy illegally in grey areas, or who have built their own homes in recent years in their own townships or group areas, often paying high prices for land because of limited amounts made available for private development in black areas.

Urban Africans under apartheid

Apartheid cities, like the whole apartheid edifice, were intended to last: hence the importance of allowing space for the future growth of white, coloured and Indian group areas. Urban Africans, however, were viewed essentially in 'Stallardist' terms, as a labour force whose workers (and their dependants) 'belonged' in the bantustans which 'grand apartheid' forged from the reserves, and as a necessary evil whose numbers should be minimized. To achieve this, the central state assumed more and more power over the lives of urban

Africans, at the expense of local autonomy. The Department of Native Affairs intervened with increasing frequency, ignoring municipal opposition to schemes such as the removal of long-established African communities in western Johannesburg (Chapter 9). In 1972 the municipalities surrendered control of urban Africans to 22 Administration Boards, which became responsible for housing, influx control and the regulation of African labour (Bekker and Humphries, 1985). Their operations were to be financed from rents, levies on employers, and profits from the liquor monopolies which they took over from the municipalities.

From 1954 onwards the policy of linking urban Africans with their respective bantustans was given substance by attempts to segregate African ethnic groups from one another within the townships (Pirie, 1984b). This made little difference to cities whose African populations came largely from adjacent areas and spoke a common language, such as Zulu in Durban or Xhosa in Port Elizabeth. The policy primarily affected the multilingual African population in the Witwatersrand townships. Whereas apartheid chose to regard whites as a single group and Africans as 10 groups, Christopher (1989b) notes that hitherto urban Africans had actually shown less tendency to segregate informally than English-speakers and Afrikaners. In practice, linguistic zoning was applied mainly in new townships, and even there segregation was much less effective than that of whites, coloureds and Indians in group areas.

The very presence of Africans in the towns became subject to formidable controls after 1948. The Illegal Squatting Act of 1951 was aimed at peri-urban squatting by Africans seeking or already in employment in adjacent towns. More fundamental was the Native Laws Amendment Act of 1952, described by Morris (1977, p. 36) as 'the most important piece of legislation in the post-war era'. It laid the basis for all state intervention to control the distribution of African labour between town and country, and between towns. Section 10 of the 1945 Act became mandatory and was extended to cover mineworkers who had previously been exempted. The Act also strengthened provision for the expulsion of Africans deemed surplus to labour requirements, and introduced the principle of efflux control by means of canalizing labour through labour bureaux in rural areas, from which permission to go to 'prescribed' (mainly urban) areas had to be obtained. Similar controls were applied to movement between prescribed areas in relation to labour demands.

Rigorous implementation of these measures, which were further strengthened in 1964, involved well over half a million prosecutions annually for transgression of the pass laws in the late 1960s. When these policies were at their peak, African population growth in the cities themselves was reduced (but not halted), and the proportion of Africans living in the bantustans increased. This was helped by the state housing policies detailed below, and in some cases by a redrawing of bantustan boundaries to include established African townships such as KwaMashu (Durban).

During the 1970s it became increasingly clear that these policies were consciously striving to create a dual labour force differentiating migrant 'outsiders' from 'insiders' with Section 10 rights, a stabilized section of the African proletariat better able to meet the growing need for more skilled African labour (Wilson, 1975; Hindson, 1987). To house the 'insiders', vast new townships were built as far as possible from white residential areas, and usually quite close to industrial areas. Whether by accident or intent, the design and location of these townships allowed them to be cordoned off in emergency, and any resistance to be smashed in the streets (Adam, 1971, p. 123).

In the late 1960s the policy shifted to one of providing accommodation in the bantustans for Africans working in adjacent towns whenever possible. This was enforced by means of General Circular no. 27 of 1967, in terms of which local authorities had to obtain government approval before initiating any African housing schemes. Extensions to townships in Pretoria, East London and elsewhere were subsequently suspended, and the city councils, acting as agents of the South African Bantu Trust, began large-scale construction programmes at Mabopane (Bophuthatswana, north of Pretoria), Mdantsane (Ciskei, near East London) and elsewhere in adjacent bantustans. Such townships accommodated not only increased African population but also those removed from existing townships such as Duncan Village (East London) and Cato Manor (Durban). So began the damming up of rural-urban migration behind artificial boundaries which has continued ever since in both townships and informal settlements, adding to the disembodied nature of the apartheid city. For cities close to bantustans, it thus becomes imperative to study the urban functional regions rather than the cities alone, and this is the criterion used in the ensuing chapters.

State responses to African urbanization

The inevitability of African urbanization to meet the changing labour needs of a more sophisticated economy has become gradually apparent to the government. Its first response was a characteristic search for incremental reform designed to optimize labour flows within a modified apartheid framework (Lemon, 1987, p. 233). The basic strategy, embodied in the report of the Riekert Commission (South Africa, 1979) and a subsequent White Paper, was to widen differentiation between 'insiders' and 'outsiders'. The former would be free to change jobs within the Administration Board area within which their 'section 10' rights were held without recourse to labour bureaux, and would be able to transfer those rights to other areas, taking their families with them, subject to the availability (as judged by officials) of jobs and housing. The intention was to enable the labour demands of the urban industrial economy to be met more efficiently by allowing a free job market for permanent residents of all races. By making the best use of those Africans who already possessed 'section 10' rights, the need to admit more to the cities could be minimized. Rural Africans would thus be pushed even more firmly to the end of the labour queue.

Attempts to translate the Riekert proposals into legislation failed, largely because of conservative drafting which failed to entrench, let alone extend, urban residence rights in the spirit of Riekert. Forces outside the state apparatus seemed to be driving towards more fundamental changes than the Riekert proposals (Hindson and Lacey, 1983, p. 104). Meanwhile, however, the legal definition of those who could potentially qualify for section 10 rights was actually narrowed by the 'independence' of Transkei, Bophuthatswana, Venda and Ciskei; children born after the 'independence' of the bantustan to which they were officially assigned could never acquire these rights, regardless of where they were born or actually lived (Budlender, 1984).

In the country as a whole, some 3.9 million Africans qualified in their own right in 1985, and a further 1.7 million qualified as dependants; they were concentrated largely in Pretoria-Witwatersrand-Vereeniging (PWV), Port Elizabeth, Bloemfontein and Kimberley (Bekker and Humphries, 1985). In Pretoria, the East Rand and Bloemfontein their numbers were steadily eroded by the use of housing as a control, as people with section 10 rights were forced to accept offers of housing in new townships outside prescribed urban areas, at Soshanguve, Ekangala and Botshabelo respectively. In Natal, changes in bantustan boundaries left few people formally qualified, although those who had been previously continued in practice to be treated as 'administrative section 10s'. In the Western Cape, a policy of coloured labour preference was followed between 1954 and 1984, and few Africans were able to qualify for section 10 rights.

In 1983, in the face of an unwelcome court judgement, the government attempted to limit the effect by amending section 10 so that the family of someone who had qualified by virtue of 10 years continuous employment with one employer could only join him if he had a house of his own rented or bought, or married quarters provided by his employer. This appeared to close the door on any continuing legal urbanization process (Duncan, 1985).

The actual machinery of influx control and labour allocation changed considerably after 1980. The centralization of recruitment in new 'employment and guidance' centres in or near African townships in larger cities increased the mobility of urban Africans and gave preference to local workseekers as Riekert had recommended. After the onset of recession in 1982, stricter controls were exercised over the entry of workers without section 10 qualifications into prescribed areas. In the bantustans, Greenberg and Giliomee (1985) document the breakdown of labour bureaux and mine recruitment as the labour market contracted and the labour surplus grew ever larger, with the result that vast areas and populations were virtually excluded from the urban labour market.

But for sheer survival, many rural households simply had to find access to urban economies. They found various means of doing so, ranging from backyard shacks in formal townships to squatting beyond easy commuting distance from urban labour markets (Mabin, 1989). By the end of the 1980s, there were an estimated 7 million informal settlers in and around the urban areas. Overcrowding of township housing reached almost unimaginable proportions in many areas; densities of 15 or more persons per four-roomed house became widespread, with even higher figures in some areas. In the metropolitan fringes, people settled on whatever land they could find: church-owned as in St Wendolin's near Durban; farms, smallholdings and vacant land in Inanda, near Durban, and around the Witwatersrand; and privately-owned small land holdings in the

Winterveld of Bophuthatswana, north of Pretoria (ibid., p. 9). Many faced eviction and relocation, often repeatedly, but for some at least their persistence was eventually rewarded as new communities became established, albeit precariously.

Others were forced to settle even further from the cities, most dramatically in the KwaNdebele bantustan. Characterized by Mabin (1989, p. 9) as 'no more than a recent and growing assemblance of land bought by the state and turned over to newly fashioned "tribal authorities" who maximise income by renting residential sites', KwaNdebele grew from minimal population to 255,000 people in 1988 (Chapter 10). Similar growth occurred in Botshabelo, near Bloemfontein, on land newly allocated to the South African Black Trust (Chapter 7). Widespread informal urbanization has also occurred in bantustan locations far beyond even the most broadly defined urban functional regions, as in the Nsikazi district of Kangwane, near Mmabatho in Bophuthatswana (Chapter 11), and in districts of Lebowa near Pietersburg. The implications of these developments for post-apartheid cities will be considered in Chapter 14.

Hopes of more fundamental reform in response to increasing awareness of the desperate struggle for survival being waged in the bantustans were seemingly dashed by the 1985 Laws on Cooperation and Development Amendment Act, which was clearly influenced by the Riekert proposals. It allowed the transfer of section 10 rights outside the prescribed areas where they were obtained, and the accumulation of such rights during 10 years of living or working in different prescribed areas. It also allowed Africans who settled in self-governing or 'independent' bantustans, or on South African Black Trust land (land destined for incorporation within the bantustans), or whose homes became part of these areas, to retain any section 10 rights they might have. Such piecemeal tinkering left influx controls *per se* untouched and the situation of rural Africans unchanged.

This 1985 legislation appeared to indicate that the government had no intention of lifting influx control in the near future. Later the same year, however, the report of a President's Council committee reflected new thinking (South Africa, 1985). African urbanization was accepted as not only inevitable but as 'an *opportunity* to utilise one of a country's greatest assets, namely its people, in such a way that the end result for all will be an improvement in the quality of life' (ibid., p. 26). The report proposed the elimination of discriminatory aspects of influx control and a shift to a positive strategy emphasizing the development role of 'orderly' urbanization. This would involve the ordering and directing of urbanization mainly by indirect incentives and disincentives but also by '*direct control measures*, mainly through existing legislation' (ibid., p. 198), including that concerning group areas, squatting and slums, health, immigration and security. In practice the continuing allocation of the greater part of the cities to whites, coloureds and Indians under group areas legislation clearly renders such an approach racially discriminatory, while the mere fact that Africans are the only ethnic group still seeking to urbanize necessarily makes them the prime objects of any constraints on urbanization.

The new approach was embodied in the Abolition of Influx Control Act of 1986, which provided for total repeal of the Natives (Urban Areas) Consolidation

Act of 1945 and the partial or entire repeal of 33 other laws. On one level these changes amount to a very significant reversal of some of the most widely criticized and strongly resisted apartheid policies including the pass laws, the insecurity arising from impermanence in the cities, migrant labour and the splitting of families. Henceforth more attention would be focused on the provision of housing, infrastructure and social services. But the new strategy still seeks to control and channel African urbanization, using material constraints of land and housing as tools in distributing labour and population, and employing the decentralization and deconcentration programmes developed under the regional development strategy which had been operative since 1982 (Tomlinson and Addleson, 1987). As Mabin (1989) argues, this strategy does much more than recognize the breakdown of 'Stallardist' controls: it actually seeks to take advantage of informal settlements on the metropolitan peripheries and further afield to fragment the African urban population in ways which facilitate indirect political control.

Within the townships, the passing of the Black Communities Development Amendment Act in 1986 made possible full home ownership and introduced procedures through which the supply of land for African housing could be increased, thus making possible the development of an African housing market. The government made clear its intention to withdraw from the direct supply of housing (which for urban Africans it had hitherto dominated) preferring to concentrate on the supply of land and bulk infrastructure. Henceforth the private sector could apply to build in development areas (land zoned for Africans in terms of the Black Communities Development Act of 1984).

The government's continuing determination to control urban growth was demonstrated by two measures in 1988. The Slums Bill attempted to give local authorities greater power to control slums and act against property owners responsible for slum conditions, limiting the number of people on the premises and setting health standards for the buildings. The Bill also gave the government power to act if a local authority failed to do so. It was strongly criticized given the overall housing shortage and the inevitability of such conditions for most of those living in them. After reference to a parliamentary standing committee, it was tabled as a white 'own affairs' Bill in 1989.

The Prevention of Illegal Squatting Amendment Act included a series of punitive measures which were attacked in similar terms. It appears that the measures are intended largely as a deterrent, which the government hopes it will not need to employ, but the Act also confers powers on farmers and local authorities who may decide to use them. More positively, the Act includes several measures allowing the provision of more land for squatting, particularly the establishment of 'informal towns', in which regulations usually governing the establishment of towns will not apply. It appears that this is intended to be a major policy instrument of 'orderly urbanization', and several such towns had already been established at the end of 1989.

Urban administration and management

Urban management in South Africa has become highly politicized. Extra-parliamentary opposition groups have identified local communities as vital units around which their struggle should be organized (Atkinson and Heymans, 1988), whereas state planners in the 1980s have seen local government as the basis for the subsequent reorganization of regional and national structures. For most of South Africa's history, local government has been in white hands. Legislation providing for the staged evolution of coloured and Indian municipal councils was embodied in the Group Areas Amendment Act of 1962. Nominated consultative committees were gradually transformed into elective management committees, but these remained advisory to white city councils, many of which consulted the committees inadequately. From 1976 onwards a series of official reports analysed the failures of the system, identifying problems of both legitimacy and financial and organizational viability. In the late 1980s Indians and coloureds largely rejected the establishment of their own municipalities, and increasingly demanded the establishment of direct representation on a non-racial basis on city councils. Cape Town City Council led the way in demanding such representation in 1985, but the situation remained unchanged in 1990.

Prior to 1977 the only official representation permitted to urban Africans was through Advisory Boards and Urban Bantu Councils, neither of which significantly influenced the decisions of white municipalities, or of the Administration Boards which took over from them in 1972. Widespread township unrest in 1976, the year of the Soweto riots, led to the establishment of Community Councils the following year. Some functions were transferred to these bodies from the Administration Boards, but they were compromised by their ambiguous relationship with the Boards, especially with regard to housing allocation which involved the councils in influx control. Polls were low in the minority of wards contested, particularly in major urban areas. In Soweto, for instance, the steady decline in the percentage poll in local elections from 32 per cent in 1968 to 14 per cent in 1974 and only 6 per cent in 1978 clearly indicated rejection of officially sanctioned structures.

The government chose to attribute this failure to lack of autonomy rather than to problems of fiscal viability and legitimacy around which hundreds of community organizations were mobilizing. It responded with the creation, in the Black Local Authorities Act of 1982, of Town and Village Councils. The former were officially regarded as fully independent Black Local Authorities (BLAs), although they were subject to much wider powers of ministerial intervention than white local authorities.

The expectation of financial self-sufficiency proved inimical to attempts by BLAs to gain legitimacy. The pressure of population numbers on buildings and services, together with low quality at the outset and three decades or more of wear and tear, demanded large capital inputs even to maintain existing housing and infrastructure. Group areas and related legislation had prevented African townships from attracting industries and minimized commercial development, thus depriving them of revenue. Given the poverty of most residents, housing is mainly sub-economic rental property; the Black Communities Development Act of 1984 did provide for the private sale of property in the townships, but financial constraints and political uncertainties had minimized the effect of this change in providing BLAs with a property tax base. Likewise income from services such as water, electricity and refuse removal could not provide a sufficient revenue base for the upgrading of infrastructure which was a necessary if not sufficient condition for improving legitimacy. Even the traditional revenue from beer halls had disappeared as a result of privatization.

Many BLAs felt that their only option was to increase rents, often substantially, but this led to resistance from communities who could ill afford such increases, especially at a time of economic recession and increasing unemployment. Rent boycotts soon became a key form of resistance to the state, and have remained so. It was the decision of the southern Transvaal BLA of Sebokeng to increase rents which triggered off the wave of violent unrest which quickly spread to much of the country and continued until brought under control by police and army action in 1986. The combination of financial and political pressures, including growing threats to councillors regarded as 'collaborators', caused the collapse of many BLAs, especially in the eastern Cape and southern Transvaal.

Just as the failure of Community Councils was attributed only to lack of autonomy, failure of BLAs was attributed solely to financial problems rather than to their fundamentally unacceptable racially defined nature. This is not untypical of the way in which the state has repeatedly grasped only that part of the nettle which it is compelled to see at a particular juncture. The government's solution to financial unviability was to include BLAs in the proposed Regional Services Councils (RSCs), hitherto planned to include only whites, coloureds and Indians in line with the tricameral parliament at national level. More immediately, the government responded to the urban crisis by using the National Security Management System as a means of implementing new policies and controls to restabilize the townships.

During 1986 joint management centres (JMCs) were established in 34 townships identified as 'high-risk' security areas, with the object of regaining control through a combined process of repression and upgrading of conditions (Boraine, 1990). The JMCs assisted the councils in various ways, and tried to present them as responsible for the upgrading schemes, but it was soon apparent that the BLAs lacked the experience to take charge of the development process. There is some evidence of selective and partial allocation of resources to benefit certain elements and divide African communities along class lines (ibid.). The JMCs largely failed in their objectives, facing growing resistance and a steady revival of civic and street structures in many areas. Many of the latter have proved able to negotiate directly with a range of state and parastatal bodies, as in Freedom Square, Bloemfontein (Chapter 7).

The Regional Services Councils Act was passed in 1985, but the first eight RSCs came into being only in 1987. The Transvaal subsequently made the fastest progress in establishing RSCs, followed more slowly by the Cape and the Orange Free State. RSCs are officially regarded as horizontal extensions of local authorities, with aims of increasing efficiency in provision of 'hard' or bulk sevices, introducing multi-racial decision-making at the third tier of government, and promotion of infrastructural development in those areas 'where the need is greatest'. In practice this means the black townships. Although the RSC Act permits the negotiated inclusion of areas across bantustan borders, this has not occurred, thus none of those in bantustan townships and squatter areas have benefited from RSC projects. In Natal, however, RSCs could have achieved little redistribution without the co-operation of KwaZulu, given the degree of interdigitation between province and bantustan. It eventually proved possible in 1989 to meet the objections of Chief Buthelezi and his KwaZulu administration, and to announce the introduction of Metropolitan Joint Service Boards instead of RSCs.

Most RSCs have in practice assumed very limited servicing functions, and with no demonstrable gain in efficiency. It seems clear that their prime objective is indirect promotion of the legitimacy of official local government structures by means of their redistributive function (Lemon, 1991). For this they are financed not by rates, which would be too politically sensitive, but from taxes on the employment and turnover of businesses, including state concerns which provide about half the income raised in this way.

RSC members are nominated by the participating authorities, including management committees. Decisions are made by a two-thirds majority, but as voting power is based on the services consumed by each local authority, this still leaves whites with a controlling interest: thus the voting power of Soweto, by far the most populous authority in the Central Witwatersrand RSC, was only 17.27 per cent in 1990. In practice, however, the (appointed) chairmen have been able to achieve consensus in most RSCs without votes being taken.

Most RSCs have given redistribution the priority intended, spending most of their levy income on upgrading infrastructure in black areas. These expenditures have undoubtedly brought tangible benefits, but they are only one element of the various moneys being channelled into the black areas, along with those from provincial, central government and private sector sources. There is little evidence that RSCs have won greater legitimacy for official local government structures; the 1988 local elections, whilst admittedly too early a test in most areas, certainly gave no indication of this (Humphries, 1989). It never seemed likely that legitimacy could be bought so easily, given their imposition from above and their dependence on official, ethnically based local government structures rather than community-based organizations and alternative civic structures whose emergence from popular action underlines their legitimacy. President de Klerk appears to have recognized this, and in 1990 he acknowledged the need for a non-racial system of local government.

The chapters which follow are thus written at a time of great change, with both the repeal of the Group Areas Act and a non-racial local government system promised for 1991. It remains to be seen what form the latter takes, and the nature of any legislation replacing group areas. The test of both will rest on the willingness of the government to negotiate the shape of the new measures rather than to impose them, so that the re-shaping of the apartheid city can truly begin.

2
CAPE TOWN
G. P. Cook

As the oldest urban area in South Africa, located in a particularly beautiful situation, it is not surprising that Cape Town should appeal to so many and be regarded as the 'mother city'. Among South African cities it is the most cosmopolitan, its character having been moulded by Dutch, Malay, French Huguenot, and British settlers in particular. Today, as in the past, the majority of residents are coloured and it is home to almost 40 per cent of South Africa's so-called coloured people. A small but rapidly increasing number of Xhosas also have their homes in Cape Town. Social interaction between groupings is limited, but homes of white and coloured residents had never been far apart until the implementation of group areas legislation which affected a greater number of people than in any other city in South Africa and has been largely responsible for the sprawl of housing over the Cape Flats.

Jan van Riebeeck did not set out to establish a town but confined himself to building a fort and laying out a garden in 1652. After 1657, settlement was limited to Hollanders and the few slaves associated with them, yet by 1660 some elements of a town plan had appeared and by 1662 the number of residents had reached 394. The estimated 50,000 Khoikhoi people in the south-west Cape lived in completely different homes and Van Riebeeck actually created a physical barrier by planting a hedge to separate them from land occupied by the new settlers with whom there had been skirmishes (Elphick and Giliomee, 1979).

The embryonic town grew slowly and by 1679 there were only 259 free burghers, 55 women and 117 children. Most of the latter must have been slaves if the recorded population of 460 including 171 slaves is correct. Khoikhoi men worked as sailors and errand boys and some presumably lived in the settlement as did any 'free blacks'. Even slow growth was discouraged when only men of capital were considered for settlement and later in 1707 when free passages were discontinued altogether. Nevertheless, in 1712 there were 250 private houses in Cape Town and the population reached 3,878 in 1714. By 1732 a well established bourgeoisie, a modest middle class and a labouring class of 'free blacks' and indigents all lived together in the small town laid out in a planned grid. Immigration was officially halted but sexual unions between settlers and locals were not uncommon and interracial marriages occurred. Notwithstanding Khoikhoi, people were specifically excluded from residing in the town unless they became mothers, in which case they were allowed to live with the fathers of their children, even if they were slaves.

When the British census (1806) was taken 21 per cent of the enumerated population lived in Cape Town where there were 1,258 private houses and a total of 16,428 people of whom 6,435 (39 per cent) were described as being of European descent. A seaside bathing resort and fishing village, renamed Woodstock, and improved roads and a military base in Wynberg set the pattern for settlement to spread along the edge of the mountain. The next 30 years saw the number of Europeans in Cape Town increase by 62 per cent. Many freed slaves also lived and worked there and by 1835 some 3,500 people were engaged in trade and industry. Over the next two decades the urban character of Cape Town became firmly established and by 1855 the town covered almost 100 hectares; there were 3,891 private houses and shops and the population had reached 25,189 (the proportion of European descent rising to 62 per cent) making it five times larger than its nearest rival, Fort Beaufort.

Improved communications laid the foundations for metropolitan growth: south with the suburban railway to Wynberg and ultimately Simonstown (established as a harbour in 1742), along the main line to the interior (Bellville established in 1861), on to the Cape Flats following a branch line and along the Atlantic coast once a road was built over Kloof Neck. Later, growth was anticipated by subdivision of land at Milnerton, Parow and Grassy Park. Towards the end of the century introduction of representative municipal government gave focus and character to the villages and stimulated growth. Variation in size and quality of homes became marked, with substantial villas on large plots on slopes, single-storey terrace houses crowded together on flat land and barracks strategically located to house African labourers brought to work on the docks (Christopher, 1984). Older parts of the town became decayed and housed a racially and ethnically mixed population under conditions of overcrowding and poverty. Indirectly this led to Cape Town having the dubious distinction of being the first city in which homes of African residents were deliberately separated from those of other citizens (Cuthbertson, 1979). An outbreak of bubonic plague was the catalyst and the central government the means by which Ndabeni location was established beyond the edge of the built-up area.

The City emerges

At the time of Union in 1904 the population was 107,000, including approximately 10,000 Africans and the densest settlement extended eastward to Salt River (Figure 2.1). Initial rapid growth of villages (the area of Bellville increased

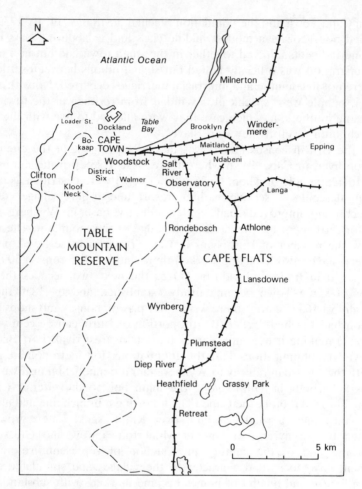

Figure 2.1   Cape Town: early development

sixfold in three years) was followed by three decades of consolidation and
incorporation starting with Woodstock in 1913 (Bickford-Smith, 1984). By 1936
most of the 350,318 people lived in a belt from Clifton on the Atlantic coast to
Plumstead on the suburban line and one in three (32 per cent white and 34 per
cent coloured) had their homes in mixed residential areas (Batson, 1947). New
predominantly coloured settlement developed along the Cape Flats branch
from Athlone to Lansdowne, and the Council laid out garden villages at
Brooklyn, Epping and Maitland, to house people in the lowest income group
according to race. By 1943 two more or less continuous stretches of urban
settlement extended south to Heideveld and east to Bellville. As it was obvious
that the various municipalities in the Peninsula would benefit if they were to
coordinate their development the first joint planning scheme was set up.

Africans were required to carry passes from 1909 onwards but could buy or
rent land anywhere in Cape Town until 1913 (and under special circumstances

until 1936). However, after 1919 African residents required exemption to live outside the location and the City Council was the first to create 'advisory boards' (when Ndabeni was handed over to them in 1925); they also introduced 'buffer zones' and racially segregated residential areas while planning a new location at Langa along garden village lines. The number of Africans employed in industries in the Western Cape increased by 534 per cent between 1933 and 1948 (Olivier, 1953). Limited segregated housing was available in Langa which became the only location after Ndabeni was cleared for industrial development in 1935. As a result an estimated 80 per cent of Africans lived in residential areas along the railway from Windermere (20,000 shanties) to Retreat, primarily in squatter shacks. In an attempt to keep them apart from the other residents the Divisional Council began planning the establishment of Nyanga beyond the edge of the built-up area in 1946.

Housing issues reflected the sanitation syndrome and attention concentrated on slums abutting the city centre (Dockland, District Six and Bokaap). The City Council recommendations to Parliament formed the basis of the Slums Act (1934) and placed the responsibility for housing in the hands of the local authority. Removals and reconstruction of some 2,000 buildings did little to ameliorate slum conditions because displaced persons were left to find their own alternative accommodation. During the 1930s attention focused on Malay people as a distinct group and the idea of demarcating a separate residential area for them (Schotsche Kloof/Bokaap) became increasingly acceptable (Truluck and Cook, 1991). However, most of the coloured community objected strongly to attempts to segregate residential areas and to setting up a department of Coloured Affairs in 1943.

Almost as many white as coloured people lived in the area between the southern suburbs and the Cape Flats branch line though one or the other group tended to dominate and form a pocket in a particular neighbourhood or village. The relative proportion of whites in the older parts of the town decreased especially between District Six and Observatory where smaller and cheaper houses were taken over by lower income groups. Although only about 20 per cent of the older medium-grade housing was occupied by coloured people, homes of residents became more differentiated for only 30 per cent of whites lived in poor quality housing as opposed to 90 per cent of coloured people (Scott, 1955). Furthermore the poorest group, an estimated 150,000 of the 482,000 residents housed themselves informally in backyards and on vacant land.

Homes torn apart

Since 1950, few homes in Cape Town have remained unaffected by government policy which radically altered the social and physical character of the city. Africans (then 10 per cent of residents) have faced a deliberate policy of exclusion. Coloured people, who made up 54 per cent of the population, were only allocated 27,950 hectares (27 per cent of land in the region) while 75,213 hectares set aside for white use included about 40,000 vacant prime residential sites and virtually all industrial land, leaving just 763 hectares for the Indians. Change in the racial patterning of the city was coupled with the loss of coloured rights (franchise in 1955 and Council membership in 1972) and with marked increases in the affluence of the Afrikaans-speaking white residents in particular. It also corresponded with unprecedented growth and by 1971 the population reached 1.1 million. A central government planning committee for the metropolitan area was established and given statutory force in 1975.

For over 30 years the full range of control measures was used against African residents of Cape Town. Compulsory barrack accommodation for migrant workers in Langa and very restricted access to limited family housing justified raids on squatter settlements and removals. For example, in only four months over 5,000 inhabitants of 'black spots' were resettled in Nyanga transit camp and 4,928 women 'endorsed out' of the city (Cole, 1986). Strict influx control and implementation of the coloured labour preference policy after 1954 made the job market very precarious and limited but also enabled business to employ Africans until sufficient coloured labour was available. Construction of houses in Guguletu adjacent to Nyanga started in 1956 but removals continued with an average of 9,000 people repatriated to Transkei and Ciskei every year for five years. Once Guguletu was completed no further family housing was built until 1980 although 11,222 Africans were evicted from Windermere, Athlone and Retreat and another 1,240 were made homeless by closing Simonstown location. Despite repeatedly returning large numbers of 'illegal' residents to the bantustans, male contract workers were increasingly brought in during the 1968−74 economic boom and housed in temporary barracks. Natural increase and immigration coupled with the housing embargo resulted in gross overcrowding of the 10,091 units available given the official population of 92,572. Although informal shelters proliferated (15,000 shacks in township backyards) squatter areas were cleared one after another until Crossroads, a purely African settlement emerged spontaneously and finally received official acknowledgement, but no protection from harassment, in mid-1976. In the face of escalating unrest, sites were set aside in existing townships for 2,575 houses in 1979 but this did little to alleviate problems facing African residents.

At the time of the group areas proclamation (30 March 1951) Goodwood, Bellville and Parow proposed that all coloured people should live south of the main line (Figure 2.2). In contrast, Cape Town City Council refused to report to the Board and only provided survey data when subpoenaed to do so. The Western Province Land Tenure Advisory Board therefore drew up its own proposals and advertised them in 1953. Besides separating whites and coloureds, specific areas were set aside for homes of Malays, Chinese and Indians. Allocating all mixed housing between the suburban and branch line as far as Wynberg to whites forced coloured people east onto the Cape Flats. By 1954 all non-contentious areas north of the railway line in the above three municipal areas as well as Pinelands, Thornton, Epping and Milnerton were zoned white. Comment was called for regarding other areas and a public meeting was held in August 1956 before the Committee of the Board made recommendations to the Minister of the Interior. The first group areas proclamations for Cape Town

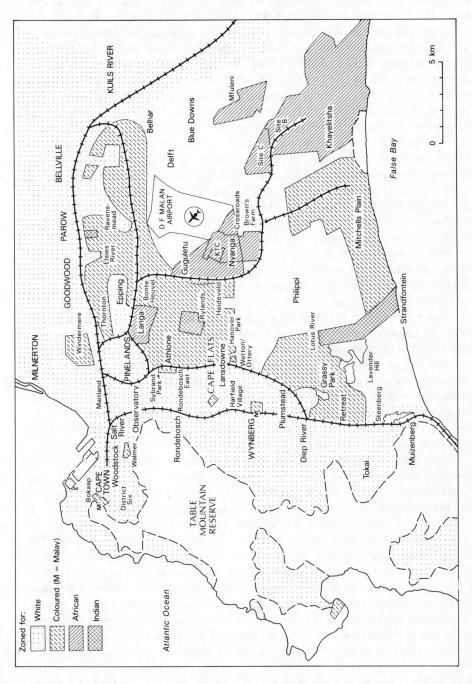

Figure 2.2 Cape Town today

were made on 5 July 1957 but most areas were not zoned until 1961. However, parts of the city centre and District Six were only proclaimed in 1965 and 1966 respectively and sections of Salt River and Woodstock remain unproclaimed.

Although occasionally efforts were made for coloured and Indian people in 'good' social and economic positions to remain in white areas, 106,177 were moved between 1957 and 1961 alone. The nature of the terrain and urban transport network exacerbated the effects of removals especially with regard to journey to work and to shops. Details of the implementation and impacts are particularly well documented by Western (1981). Yet some areas between the lines never became completely white and the 1985 census showed that 9 per cent of coloureds still lived in non-coloured designated areas. For example, in 1969 Lansdowne residents were given 15 years to move out and after 1984 coloured families remaining were increasingly pressurized. Nevertheless, others rented or bought houses in close corporation property deals and in June 1990 the Wetton/Ottery border was declared a free settlement area. The character of Rondebosch East remained such that the first non-racial civic association serving a mixed community in a white group area emerged there in 1990. In contrast, some families moving into officially white neighbourhoods, even at the upper end of the market are harassed or made to feel threatened as in Goodwood where the 1989 Christmas circular included a request to report 'trespassers' of other races.

District Six, one of the original wards of the city, once home to a dynamic, primarily coloured community was zoned white and remains a focus of protest and a symbol of group area removals. In contrast its near neighbours, Walmer and Bokaap, are centrally located residential enclaves for Cape Malays. Despite general physical neglect and a variety of land uses District Six was primarily residential with a social and ethnic mix including homeowners, tenants and squatters who occupied the 6,122 houses (71 per cent owned by whites, 18 per cent by coloureds and 11 per cent by Indians). Property transactions were frozen in January 1965 and after proclamation 13 months later the 33,446 coloured people officially resident (together with an estimated 5,000 illegal tenants) were given a year in which to leave. However, between 55,000 and 65,000 people were actually relocated, most before 1978 when it was renamed Zonnebloem. The cost of compensation for buildings alone was R30 million. In December 1979 there were 2,000 families left and the first few whites moved into renovated flats in 1981. Clearing was completed in 1982 when the last Indian families were evicted from 20 of the houses left standing (Hart, 1988). Neighbours became scattered, many were resettled in Belhar and Hanover Park, some who could afford to went to Mitchell's Plain or to Walmer if they were Cape Malay and Indians had to settle in Rylands. Much of the 17 hectare site remains vacant except for the Technikon and Oriental Shopping Plaza and there are fewer than 4,000 residents (many Portuguese) despite government involvement in upgrading and building houses, providing cheap loans for home-owners and accommodation for state employees. The recommendation of the President's Council that District Six should be for coloured residents was rejected, though in 1983 a portion (20 per cent) in the east was reproclaimed coloured. Thirty-four houses were built there in 1986 but the project remains controversial and is incomplete. Only three coloured families have bought properties. In 1989 part of District Six became one of the first free settlement areas to be proclaimed but its future use remains uncertain in the face of strong opposition to the land being used for 'elitist' dwellings and quasi-state institutions given its history, the acute low-cost housing shortage, and the call to scrap group areas in Cape Town as a whole.

Given that many who lost their homes were unable to afford accommodation and became dependent on local authority housing, both the city and the Divisional Council, with a total of only 10,551 and 4,000 units respectively, embarked on large-scale housing projects to provide the 99,576 units in existence today. One of the first schemes was Bonteheuwel, located near the industrial area and served by two rail links. However, as the housing problem escalated in the late 1960s and early 1970s so new estates were located on less appropriate sites further out on the Cape Flats (Younge, 1982). Only in the case of Bokaap were Cape Malay residents allowed to remain in houses the Council restored and rebuilt behind facades between 1969 and 1976. Money for housing projects had to be obtained from loans under the provision of the Housing Act (1966) and at first 50 per cent and later 25 per cent of all new units were allocated to displaced families. Private bodies such as the Cape Flats Distress Association and Citizens Housing League were also involved to a limited extent.

The largest single development was that of Mitchell's Plain, the building of which began in 1974 on 3,100 hectares 27 kilometres south-east of the city. Choice of site was somewhat controversial as it was far from existing residential areas and away from the axis of development, had very poor communication links with the city, lay on unconsolidated dune lands and there was some question about the land deals. Nevertheless almost a quarter of a million people now live in about 40,000 homes at an average density of 45 houses per hectare. Improvement of road links and completion of the railway may have reinforced the dormitory status of Mitchell's Plain, but development of commercial, social and medical facilities coupled with the lack of alternative housing opportunities have encouraged growth. Although originally aimed at home ownership the fact that Council tenants occupy a third of the houses is reflected in the 1985 average head of household income of R700–800 per month. The mixture of house size and quality has led to a degree of differentiation between, for example, the north-west where most larger privately built houses are located and the low-income problem areas in the south-east.

In 1975 the government released plans to create a completely self-contained city, Atlantis, 45 kilometres north of Cape Town, for 500,000 people. The intention was to encourage industrial growth and provide jobs for coloureds while acting as a deconcentration point for Cape Town and a link with probable development at Saldanha. Building has hardly extended beyond the original south-east sector and until 1985 housing was restricted to people employed in Atlantis. Despite incentives only 128 factories occupying 21 per cent of the available industrial land had opened by 1987 and each of the 14,800 jobs are estimated to have cost between R66,000 and R80,000 to create. Clothing and textile industries predominate (23 per cent of factories and 42 per cent of the jobs) and 20 per cent of concerns employ less than 10 persons: the multiplier effect is minimimal. Today 55 per cent of the economically active residents commute to other centres and Atlantis has become a low-income satellite in which an average of 31.9 per cent of household income is spent on rent, over 51 per cent of the 7,750 households are in arrears and there are serious social problems (City of Cape Town, 1988).

Introduction of group areas had very little direct impact on white residents of Cape Town for originally only 11 families were required to move. However, the less prosperous were able to buy houses vacated by coloured families especially in parts of Lansdowne, Observatory and Windermere or to move into cheaper rented accommodation in Woodstock for example. The changes also worked to the advantage of developers who bought blocks of vacated cottages and renovated and adapted them while capitalizing on their Cape character as in Harfield Village and parts of Wynberg. In the case of the Loader Street area profitability was enhanced by the combination of good views and convenient location near the centre of town. Although 38,000 un-developed single residential plots were available for white occupation in the metropolitan area in 1977 both local and divisional councils opened up townships on the slopes of the Tygerberg and peninsula mountain chain where affluent whites were keen to settle on large plots (Zietsman, 1980). New medium-priced white housing developments have encouraged squatter clearance as in southern Tokai. Widespread subdivision of white residential land concentrates in northern Milnerton. Here developers respond to insecurity generated among white residents by erecting compact, secure homes as far away as possible from the Cape Flats. Since non-residents have been permitted to invest in property and the value of the rand has declined, demand for large homes, especially those with spectacular views over the Atlantic, has steadily increased and property prices at the top end of the market have escalated. However, in the face of the falling white birth rate and an ageing population there is an oversupply of houses especially starter homes, reflected in an average of only 2.93 persons per house in white residential areas.

A city in crisis

By 1980 the Cape Metropolitan Area was the second largest in the country with the total population officially 1.8 million. According to official sources 28,829 coloured, 1,465 Indian and 248 white families had been moved and only 2,261 families remained to spoil the neatly segregated pattern which the City Council decried in its Open City initiative (Dewar, 1980). Less than 6 per cent of the coloured community still lived in central areas; the remainder were concentrated on the Cape Flats, 52 per cent of them in the southern suburbs. Employment opportunities focused on the centre (46 per cent) and central periphery (26 per cent) and with almost all the working-class population more than 15 kilometres out of town the longest journey-to-work became associated with ethnicity and low economic status (van der Merwe, 1985). On average 83.2 per cent of household heads in coloured and African suburbs, as opposed to 5.5 per cent in white areas, had no car in 1989.

The economically active population increased by 2.7 per cent from 125,000 in 1960 to 211,000 in 1980, but the overall position became increasingly less satisfactory. Although still in second position nationally, the relative contribution to the GDP remained based on light industry but dropped from 11.6 per cent in 1960 to 10.4 per cent in 1980. Over the same period the contribution of manufacturing to GDP declined from 15 per cent to 10.4 per cent as the metropolitan area fell to third place in the national economy. This decline is significant because manufacturing provides 25 per cent of jobs in the metropolitan area and employs 30 per cent of coloured workers. Although these employees are relatively skilled and generally have a moderate level of education their opportunities are restricted and with the scrapping of the coloured labour preference policy in 1984 competition for jobs at the lower levels is growing. There is limited room for industrial expansion: the vacant third of zoned industrial land is expected to be used up within 10 years even at the present slow rate of industrial growth. The tertiary sector (especially finance and government) has expanded but already predominates to a greater extent than in any other metropolitan area. Unemployment levels among registered workers have undergone a fourfold increase since 1980 and are conservatively estimated at between 11 and 18 per cent (Thomas, 1990). Furthermore, 15 per cent of the labour force are estimated to be employed in the informal sector where under-employment is high and renumeration generally low and variable.

The situation has very serious implications for housing. In 1983 it was estimated that just 6.3 per cent of coloured people earned over R620 per month and that only persons with a monthly income of more than R800 could afford housing in the private sector (City Engineer's Department, 1983). Even subsidized purchases are limited, for despite 11,772 applications in 1983, by 1988 only 31 per cent of City and 46 per cent of Regional Service Council houses had been sold. The demand for housing among lower-income groups is illustrated by the City Council waiting list of 1983 which had 650 white, 286 Indian and 20,380 coloured people requiring rented accommodation. Furthermore, over 22 per cent of coloureds on the list earned less than R200 per month and a mere four per cent between R650 and R800. Yet 98 housing units were built for white occupation and only 4,494 for coloured people in 1982. Land constraints also inhibit housing schemes because 10,839 residential plots in the Peninsula are available for whites, but only 3,646 for coloureds and 242 for Indians (South Africa, 1988a). The deteriorating situation is reflected in 70,000 applications for housing assistance, 80 per cent from families with single breadwinners who earned less than R600 per month in 1989.

Although the average proportion of income spent on rents and electricity for all income groups was 19.6 per cent in 1983 it is far higher, and rises to over 33 per cent, for people who live in economic and sub-economic housing schemes. Not surprisingly therefore by mid-1986 almost 48 per cent of tenants in units rented from the Cape Town City Council were in arrears. The situation is unlikely to improve for a survey in Macassar showed that even with decreases of 90 per cent in rent, 23 per cent of occupants of the cheapest housing still would not be able to pay for basic necessities. Evictions (558 for rent arrears in Grassy Park in 1989), long waiting lists and attempts to share the rent burden have led to gross overcrowding and to more than one household occupying a property. Almost two-fifths of council flats are estimated to be occupied by two families and at least one house in five has backyard dwellers. There are many recorded cases where families have to live in the rural area while the breadwinner commutes, sometimes on a weekly basis, and sleeps in a car or derelict building while away from home. On the broader front 86 per cent of all sub-economic, 71 per cent of economic and 47 per cent of private houses in coloured residential areas in 1985 were overcrowded. Home owners have responded to the pressure in various ways. Rooms have been added on or extended, garages have been converted into flatlets and timber frame shelters, some of them lean-to, have been erected on plots particularly in older residential areas such as Athlone and Heideveld, where plot sizes are relatively large. Local authorities have increasingly turned a blind eye to these contraventions of building regulations.

There are a limited number of areas where coloured people are found openly squatting but numbers involved are nevertheless large. In 1984 spontaneous settlements scattered in Elsies River, Philippi, Lotus River, Grassy Park, Retreat and Ravensmead housed 2,502 families. These residents are given priority in housing schemes. For example families in 174 shacks south of the open area to be developed in Lavender Hill East were each allocated a 10 square metre core with tap, waterborne sewerage and prefabricated wood and iron panels with which to build homes at rentals of R1.88 per month. Nevertheless the impact is limited and 213,000 people in the region were officially regarded as requiring formal housing when the radical structural change associated with creating the Regional Services Council occurred in 1985 (Todes, Watson and Wilkinson, 1987). The City Council took the opportunity of expressing its commitment to an open city in which 'the only equitable and satisfactory form of local government is for full participation by all citizens irrespective of race, colour or creed' (City of Cape Town, 1986, p. 75). It also argued for preservation of municipal councils and opposed fragmentation of the metropolitan area. Simultaneously, the council went ahead with a variety of alternative strategies aimed at reducing the housing problem, confirming the city's reputation for innovative approaches. Among these were bean-bag housing schemes at Lavender Hill which have the dual aim of cutting the waiting list while providing jobs. In 1986 the Urban Foundation initiated building of 391 self-help units. Nevertheless when the Regional Services Council took over in August 1987 it was faced with an estimated housing shortage of somewhere between 29,200 and 45,000 units (Technical Management Services, 1988).

During 1983–4 the City Council attempted to address the coloured housing problem by raising a loan to create another new town, Blue Downs, on a 4,000 hectare site near Kuilsriver. The plans were passed to the House of Representatives which finalized the structure plan before building started late in 1986. The scheme includes a CBD and 'activity axes' along which home-based small businesses can be located. As yet no date has been given for completion of the rail link. Although the intention is not to create another dormitory town only 130 hectares is zoned for industry in relatively small complexes. The anomaly is explained by proximity (15 km) to existing industrial areas and a need to cater for emerging local concerns. Plots range in size from 300–350 square metres and except for a limited number set aside for private development by individuals, all building has been contracted out. Using the neighbourhood unit concept six developers working with co-ordinating committees aim to create a total environment satisfactory to all. Attempts have been made to keep costs down and to cut out middlemen but the development has not been without criticism especially with regard to escalating prices associated with rapidly increasing land costs (now controlled) and false economies practised by developers. Properties average R55,000 and purchase requires a deposit and a regular household income of at least R960.00 per month. As only 38 per cent of coloured households earn more than this amount applicants are restricted. Mortgage repayments on the cheapest two bedroomed house make it impossible for a family at the margin to do more than subsist in the short term at present wage levels and arrears and repossessions are therefore inevitable (1,095 in three weeks in November 1989). With this in mind, the Delft project was initiated, starting with 2,000 'affordable units' on sites adjacent to the airport. Here the breadwinner only need earn over R400, be on the official waiting list and be nominated before becoming eligible for a state loan at a low fixed rate. Plots are 2–300 square metres and cost R8–12,000 including transfer, service connection, fencing and landscaping. In addition education, health, social, sport and community facilities are all provided at the outset and 80 hectares have been set aside for industrial development. Residents are directly involved at all stages and are supported by a social planner/domestic consultant in an interactive three-tier communication system. Participants may employ their own builder, choose and adapt one of the 56 prototype houses or join the assisted self-help scheme under which materials and technical supervision are available from the project contractor. Although the latter option does reduce costs, prices are over R15,000 and more commonly R20–25,000, still well beyond the range of the poorer group.

The official announcement in 1983 that Africans legally resident in Metropolitan Cape Town would be housed in a new town, Khayelitsha, to be constructed on the south-east edge of the city seemed a reversal of earlier policy (Cook, 1986). Plans for the 30,000 core houses did not at first include land rights and were linked to clearance of existing townships and squatter areas. Levels of unrest were such that the end of 1984 saw 99-year leasehold introduced and site and service plots developed on the periphery (site C) intended for sports and educational uses in Khayelitsha. Two months later permission was given for Africans officially resident in the three townships to purchase their homes (leasehold) if they wished. In 1985, when the African population reached 350,000 (16 per cent of Cape Town's total), the Urban Foundation recorded 20,108 formal housing units for the community, 83 per cent of which were rented. There were also 25,000 hostel beds, a figure which may mislead, for a survey has shown that typically a family unit including children occupies each bed in a hostel block (Dewar and Watson, 1990). Thus as many as 20 families share rudimentary barrack type toilet and cooking facilities. There were also between 42,930 and 44,639 informal dwellings with African residents. Since then the size of the African population has increased dramatically and housing has become predominantly and increasingly informal in nature despite the original concept of Khayelitsha and involvement of business and developers in providing accommodation in township buffer zones. In Khayelitsha sites B and C are designated for informal settlement but recently virtually all open land has been covered by dense clusters of squatter shacks. Some running water and sewerage facilities are provided in the former areas but not in the latter. Other Africans have homes in demarcated squatter camps of Crossroads, KTC and Nyanga East as well as the tent towns in New Crossroads which are also provided with some services (Cole, 1987). Finally, within the townships themselves, large numbers of informal units are constructed in backyards (92 per cent of surveyed township houses) and remain by tacit agreement with officials. Nevertheless housing densities are high for 62 per cent of formal units have more than six persons in two or three rooms, more than half the small backyard shacks house three or more people and even demarcated informal units have over two persons per room. Although some residents in the formal sector have purchased their homes (700 out of a possible 5,000 in Khayelitsha and 1,700 of 14,000 in the other townships in 1988) very few even contemplate this for it is estimated that 42.7 per cent of African families in Cape Town earn too little to be able to make any payment at all towards housing costs.

In the past decade, housing problems have reached crisis proportions as evidenced by gross inequalities (Table 2.1), the low and steadily deteriorating quality of existing stock, the unavailability of finance to support present structures let alone provide new ones, and the inability of most residents to meet even minimal charges. Inadequacies of the housing environment impact on all aspects of the people's lives and are reflected in escalating health, educational and social problems (Rip and Hunter, 1990). Despite deployment of resources into housing programmes, the backlog increases steadily, especially as 1989 saw the lowest number of houses built since 1956. For the coloured community it is between 50,000 and 82,000 units and can be expected to rise to 108,000 by

Table 2.1 Metropolitan housing inequality

Group	Estimated population (%)	Units (no.)	(%)	Families on waiting lists	Rent arrears (%)	Average units per hectare	% units <3 rooms	% units >5 people	Persons per room
White	23	197,671	48	–	–	8	<1	–	0.8
Coloured	48	195,114[1]	47	65,048	49	19	87	72	2.3
African	27	20,108[2]	5	90,000	85	20	85	89	2.6

Notes
[1] Plus 2,916 informal units
[2] Plus 43,785 informal units

2000, even without rehousing shack dwellers. For Africans it lies between 33,000 and 41,000 (or 80,000 if official shack dwellers are included) and will reach 125–165,000 units at the turn of the century. With present approaches there is neither land nor money to provide homes for all.

Futures

Although a variety of scenarios incorporating removal of group areas are contemplated for the future of Cape Town (South Africa, 1988a; Zille, 1989; Thomas 1990) of necessity these are based on present trends and past experience under white control and in the light of West European mores. Development of the city in the twenty-first century will be dominated by people with different priorities, values and requirements. Whatever trajectory is selected, the built form of Cape Town will remain as the frame within which population will double to between 4.2 and 4.8 million by the year 2000 and will double again in less than 30 years. The city's future will reflect the degree to which limitations are appreciated and existing advantages can be capitalized on if an urban environment that maximizes opportunities for every future resident is to be created.

In Cape Town where 28,000 new households are formed each year and between 300,000 and 500,000 housing units will be required by the turn of the century, problems with people's homes are one facet of those affecting the system as a whole, in particular rapidly increasing population numbers coupled with an economic slowdown. This inter-relationship must inevitably affect approaches to housing for 'only when the general problems of socio-economic development of this area have been tackled successfully will the very poorest people also be able to afford decent housing for their families' (City of Cape Town, 1987). Prerequisites are fundamental changes in attitudes to land utilization and urban development in general coupled with acceptance of the fact that the major resource of greater Cape Town lies in its people.

Even if one disregards earlier growth patterns, more than 30 years of group area implementation has structured the city so that the population potential is concentrated east of the Cape Flats line (Figure 2.3). In particular, a high density axis extends in a north-south direction from Kraaifontein to Macassar where more than half the metropolitan population live. Expansion of working class housing from Blue Downs north to Kuils River together with completion of Khayelitsha and growth of nearby Mfuleni will result in high concentrations of low income earners in an outer arc. Here the business and employment potential of large numbers coupled with newer road and rail links between existing routes to the interior is likely to attract entrepreneurs in both formal and informal sectors: the planned shopping complex at the Khayelitsha interchange is a forerunner.

Given expected rates of population growth and new family formation in this eastern peripheral arc, provision of shelter and related services is of major concern. Even if the annual increase in number of Africans drops from the present 14 per cent to the expected 6 per cent, large numbers and a serious

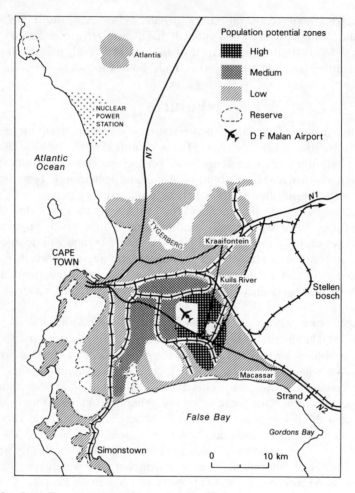

Figure 2.3   Cape Town in regional perspective

shortage of capital make present responses to housing both impractical and
inappropriate. Neither are they acceptable to the majority. What will be
important is an environment that capitalizes on the expertise of local residents,
develops house building skills, facilitates construction and either makes use of
existing infrastructures or creates new ones. Man not technology can be expected
to dominate and inputs at an individual level will be both necessary and
effective. Space and performance criteria rather than building and physical
standards may help encourage improvement of existing housing, whether informal as in Sites B and C and Brown's Farm, or formal as in sub-economic
schemes in Khayelitsha and Blue Downs. Evolution of integrated natural living
environments reflecting community priorities and requirements may facilitate
the urbanization process of African migrants and reduce stress in the poorest
members of Cape Town's population. Related strategies include overcoming

shortages of industrial land by highly selective and limited concentration of industry in already established areas in the outer arc and encouraging a land use mix appropriate for small-scale business and industrial development. Problems of encroachment on nature reserves (already being experienced at Driftsands Reserve) and on limited high quality agricultural land may in part be relieved by allowing eastward expansion of Khayelitsha on to land presently used by the Department of Defence. Ultimate coalescence with the 2,000-plot extension to Macassar will result in the built-up area fringing the whole of the False Bay coast.

Pressure on existing housing is greatest in the area between the Cape Flats line and the low-income fringe. Here demand has artificially inflated prices, multiple occupancy is the norm and squatter shacks abut on quality homes – all features that exacerbate poverty and inequalities. The challenge is to alter the present inefficient structure and rigid pattern of land use to create a truly enabling environment. An unconventional approach to town planning coupled with practical responses to grass-roots ideas and demands is essential. Encouragement of land use mixes, infill and high-density housing together with relocation and linear development of social and recreational facilities on vacant lots near high demand areas should help conserve land and increase accessibility. In this regard any decrease in separation between residence and work-place for example would reduce both time and money spent on commuting and free resources for more productive purposes, in particular improving existing housing stock. Upgrading of roads, footpaths, street-lighting and drainage already begun in 45 communities, together with the opening up of convenient residential areas like Rylands, the already mixed suburbs of Lansdowne, Rondebosch East and Sybrandt Park and the remaining Cape Flats suburbs should increase residential opportunities. Responses to the challenge presented by limited land for the 12,100 homes required each year for these residents include intensification, utilization of existing agricultural and mining land adjacent to Mitchell's Plain and northward extensions along the route towards Atlantis.

In contrast, the inner arc from the Atlantic seaboard, along the mountain fringe to the False Bay coast has the maximum potential for sub-division and intensification of residential use. It is estimated that the number of homes west of the mountain chain could be doubled. Areas of high accessibility along the suburban line and near the central city are likely to attract numbers of people presently living on the Cape Flats while the most upwardly mobile residents compete for more prestigious, low density accommodation against the demands of those living in the outer arc where the pressure on land is greatest.

Faced with a huge housing backlog together with an escalating demand, one of the problems is how to balance needs of local communities against the requirements of the total system and to achieve the holistic view so essential for effective operation. Solution of the housing problem in Cape Town requires that the constraints of a lagging economy, a serious shortage of water and insufficient land for urban development should all be overcome simultaneously. The more jobs that become available, and 200 new ones are required each working day to keep pace with growth, the greater the potential of the metropolitan system and the more likely the growth of the housing sector. Yet neither can occur unless a way is found of overcoming the limitation imposed by a water supply which after exploitation of all additional sources, will not meet basic requirements shortly after the turn of the century. Finally a means must be found to use all available land effectively if there is to be sufficient space for residential, industrial and service provision. These constraints in turn are dependent on and interact with the structure, control and organization of the metropolitan area as a whole. From being paternal in outlook and protective in approach a shift of focus to include local organizations, such as the recently formed federation of municipal associations of Cape Town, and to take cognizance of grass-roots inputs is required. In this regard political, social and economic security are important factors in fostering an atmosphere of goodwill and mutual support. Problems regarding the financial base and how to achieve redistribution require solution. Refinement of ideas incorporating land tax systems and allocation according to need are required if housing at all levels is to benefit and disparities between the two worlds of the mountain slopes and the outer fringe are to decrease.

Given the prevailing situation, the priority in Cape Town is stimulation of development and creation of jobs rather than provision of housing for a burgeoning population. A conventional approach that relies on large-scale building of houses according to specified norms in prescribed areas and using public monies is neither practical nor acceptable and radical change is required. This involves responsive and positive reactions to innovative ideas, progressive thinking, being prepared to encourage individual endeavour and above all the creation of enabling social and physical environments that will break the legacy of the past and achieve a city characterized by homes for all.

3
PORT ELIZABETH
A. J. Christopher

Port Elizabeth has been one of the principal cities where the foundations of both apartheid and the opposition to racial discrimination in South Africa were laid. The earliest designated, separate, residential area or location for blacks in the country was established in 1834 in Port Elizabeth. Yet the city has been one of the major centres of opposition to attempts to coerce, disenfranchise and dispossess African citizens (Riordan, 1988a). As such the development of the present pattern of ethnically segregated residential areas is a complex and contentious subject.

Before group areas

Port Elizabeth was laid out in 1815 as a British colonial port to handle the import and export trade of the eastern Cape. The first settlement was built on a narrow strip of land between an old marine cliff and the sea. Later suburbs were developed across the higher elevations, as differentiation on the basis of economic class amongst the predominantly white population assumed importance. By the 1850s the sprawl of suburbs, smallholdings and industrial zones which characterize the modern city had begun to take shape. Unlike Singapore, founded four years later, no provision was made for housing the various components of the heterogeneous population in separate districts. However, in 1834 the London Missionary Society established a separate location for the black population under its pastoral care at the port (Christopher, 1987a). It should be noted that the Society had already established a separate village for black persons in 1803 at Bethelsdorp some 15 kilometres to the north-west of the town.

The site chosen for the location was approximately 0.8 kilometres from the town centre and separated from it by the cemeteries and open land. The influx of Africans into the town in the ensuing decades was such that the municipality established a new location, the Native Strangers Location, adjacent to the original in 1855. As the name suggests the location was designed exclusively for Xhosa and Mfengu workers, who were deemed to be temporary residents and thus only offered 30-year leases on the plots in the expectation that at expiry they would return from whence they had come. Africans living elsewhere on the open common lands around the town were required to reside in the location and measures were taken to enforce the rules.

The town grew rapidly in the second half of the nineteenth century as the trade of the eastern Cape expanded (Table 3.1). Extensive new suburbs for the white and coloured population were built. It should be noted that although the franchise in the colonial period was nominally colour-blind, there was a notable correlation between skin colour and economic status. Poorer, working-class suburbs tended to be darker skinned than the wealthier, middle-class suburbs. Thus in general the higher the elevation and the further from the city centre and port the whiter the population tended to be. However, Indian and Chinese traders, who began to arrive late in the nineteenth century, were distributed throughout the town, except in the African locations from which they were excluded and the Central Business District which was too expensive.

There was a parallel establishment of new African locations beyond the urban fringe. The majority of the locations were administered by the munici-pality and strictly controlled, but one, Gubb's, was privately run and its inhabitants were allowed a wide measure of freedom to maintain aspects of traditional African society, which was expressed particularly in building styles. Thus by 1900, Gubb's Location housed more Africans than all the others combined. Demands by the white population for the removal of the older, inner, locations grew as new suburbs surrounded them and whites were therefore brought into closer residential proximity with Africans. In 1901 bubonic plague

Table 3.1 Population of Port Elizabeth 1855–1990

Year	Whites (000s)	Coloureds (000s)	Asians[2] (000s)	Africans (000s)	Total (000s)
1855	4	1	–	1	5
1875	9	2	–	2	13
1891	14	6	–	4	24
1911	20	12	1	8	41
1936	54	28	2	30	114
1951	80	46	4	71	200
1960	95	68	4	121	288
1970	120	95	5	167	387
1980	139	121	7	252	519
1990[1]	145	150	8	500	803

Notes
NB. Totals may not agree due to rounding.
[1] Estimate, other figures based on census (1980 on revised figures to allow for undercount).
[2] Asians include Indians and Chinese.

broke out in Gubb's Location, and the municipality took the opportunity to remove all the existing African locations and rehouse the population at New Brighton, 8 kilometres from the town centre (Figure 3.1). Owing to the delay in implementing this proposal possibly half the people involved moved to freehold land at Korsten, beyond the existing municipal boundary. However, at Union in 1910 half the African population of Port Elizabeth lived in formal locations or migrant workers barracks.

In the Union period the pressures towards greater racial residential separation intensified under the prevailing concept of pragmatic segregationism (Christopher, 1987b). However, the proportion of the African population living in the locations changed little until the 1950s. The political position of

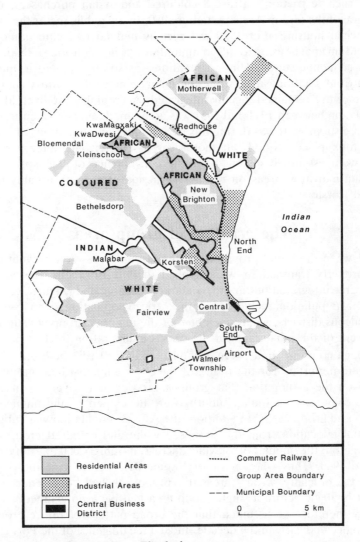

Figure 3.1 Ethnic zoning in Port Elizabeth

the Africans deteriorated as they were totally disenfranchised and no longer able to purchase land freely after 1936. Thus increasing numbers were housed by their employers in domestic quarters and company barracks within areas deemed to be part of the white town, and densities on land that did belong to Africans increased.

The first half of the twentieth century witnessed the initial moves to differentiate between the various non-African groups in the remainder of the town. In the private housing sector this was attained through the insertion of racially restrictive clauses in title deeds of new suburban developments. The majority of such deeds restricted ownership and occupation rights to whites only, although the housing of domestic servants was accepted. Accordingly those suburbs without such restrictions attracted coloured and Asian purchasers, until the imposition of the Group Areas Act after 1950. In the public sector the allocation of municipal housing after 1920 was racially defined to create homogeneous estates. Municipal housing for all communities was in a uniform style, dependent only on economic status. Thus segregation was strictly applied in most areas laid out from 1900 onwards, while the inner, nineteenth century suburbs and the town centre remained racially integrated. Accordingly the fivefold increase in population between 1911 and 1951 was accommodated almost entirely within segregated suburbs. It was during this period that the foundations of industrialization took place, notably in motor car assembly. The inter-war housing estates were essentially planned for the convenience of the workers in the expanding industrial areas on the northern side of the city notably in North End and Korsten.

The creation of group areas

By 1951 when the initial provisions of the Group Areas Act preventing inter-group property transactions were applied to Port Elizabeth, the city was in some degree integrated but in others markedly segregated (Figure 3.2). A third of the total population, including half the African population, lived in essentially homogeneous districts, where members of the majority group constituted over 90 per cent of the inhabitants of the census enumeration tract.

Although no group areas were promulgated in Port Elizabeth until 1961, the 1950s were important for the future of the city. Municipal housing plans were pursued on the assumption that group areas were effectively in place. Thus separate coloured and Indian suburbs were developed by the municipality in the proposed group areas. In addition, the African population was subjected to substantial relocation from the proposed white and coloured group areas to New Brighton, the population of which increased from 35,000 to 97,000 between 1951 and 1960. Thus whereas in 1951 some 31.3 per cent of the population lived in the officially 'wrong' area, this proportion had been reduced to 16.2 per cent in 1960, before a single group area had been proclaimed.

It may therefore be assumed that the broad pattern of future group areas had been decided early in the decade. Indeed the diligence of the Port Elizabeth municipality in enforcing racial zoning was commended by government ministers

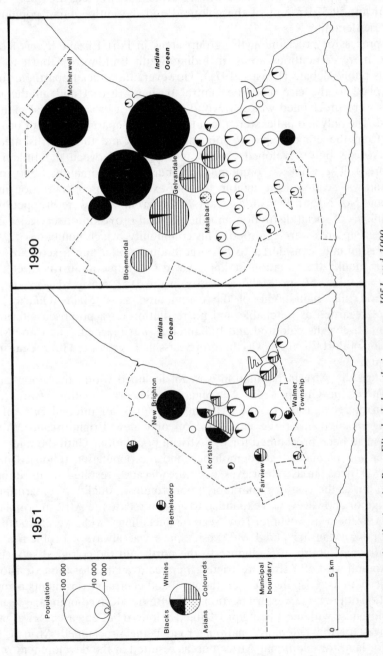

Figure 3.2 Population distribution in Port Elizabeth by ethnic groups, 1951 and 1990

as early as 1950. Assuming that the broad outlines of future group areas had been determined by 1950 then at the time of the 1951 census only 1.1 per cent of the white population was faced with resettlement, whereas 43.1 per cent of the African, 59.5 per cent of the coloured and the entire Asian populations faced resettlement.

The process of proclaiming the group areas in Port Elizabeth was lengthy and the more contentious areas, including South End, were subject to acrimonious public debate (Davies, 1971). However, the official apartheid model was applied to the city, and the central business district and all the inner suburbs were proclaimed white. Beyond the centre a broad sectoral plan was adopted. The only anomalies were the survival of the separate Walmer Location, dating from the era of separate municipal status, and the establishment of South Africa's only developed Chinese group area between the Indian and white areas. The Chinese community numbered approximately 1,500 in the 1950s and was scattered throughout the city except for a concentration in the centre and North End. Plans for a Cape Malay group area were dropped when the problems of social disruption and the physical provision of services for yet another group area were explored. This community, which numbered 2,911 in 1951, was initially considered to be large enough to sustain a separate area. However, unlike the situation in Cape Town no enumeration tract, let alone suburb, recorded a Cape Malay majority in the 1951 census. The concept of South End, the main home of the community, as a 'Malay Quarter' was therefore regarded as untenable and no resolution on a proposed alternative situated between the coloured and Indian areas was reached. The Cape Malay population of Port Elizabeth was therefore classified as coloured for the purposes of the Act.

Although the African population was not included within the ambit of the provisions of the Group Areas Act regarding the proclamation of areas, the provision and extension of the African areas was an integral part of the emerging racial plan of the city. In 1950 only new Brighton and Walmer locations had been proclaimed for the African population. Until the mid-1980s the latter was planned to be demolished and the population rehoused in the northern African areas. Reprieves over the decades resulted in its ultimate survival, but at the cost of remarkably poor, rundown quality of environment. New Brighton, however, was expanded to accommodate the African population of Port Elizabeth as a whole. New sections including Zwide were proclaimed and later KwaMagxaki and KwaDwesi. Space was always a problem as the detached white suburb of Redhouse to the north and the coloured suburbs to the south and west effectively restricted the area available for the extension of the African areas. Dispute between the local and central governments over the designation of the large western portion of the Bethelsdorp commonage resulted in a compromise with most being proclaimed coloured but one area, KwaDwesi, to the west of the Uitenhage national road being proclaimed African. The search for land for additional African areas resulted in the development of the detached area of Motherwell to the north of the Swartkops River in 1982. This suburb is capable of extensive expansion over low quality farmland. The present African areas are therefore separated from one another by blocks of land proclaimed for other groups, and are administered by three different local authorities.

The deleterious process of fitting the population into the group and development areas drawn on the map involved a long process in which some persons were moved more than once as the various stages were implemented. Provision of housing clearly was a major constraint on the programme as expropriation and eviction were complemented by house building and resettlement. Thus population exporting districts had to be matched by reception districts (Nel, 1988).

Change in the city

In the central suburbs, sections were razed after everyone had been removed. The largest such zone in Port Elizabeth was South End where virtually the entire suburb housing some 12,000 inhabitants was demolished. In North End rows of houses, even entire street blocks, owned or occupied by Asians and coloureds were destroyed. In the proclaimed coloured section of Korsten, all the properties expropriated were demolished. Accordingly, in a suburb in which the coloured, African, Indian and Chinese inhabitants had been markedly integrated the resulting pattern is chaotic. In other areas including much of the central zone, few properties were razed and the expropriated or non-conforming properties were sold to white persons. Municipal housing estates occupied by tenants of the 'wrong' group were subject to a policy of house exchange. In general this meant the rehousing of coloured tenants in the new coloured group areas and the occupation of the estate by whites. However, in a number of cases the demand from whites for municipal housing was insufficient and the estate was demolished and the land left vacant. The last such removal and demolition took place at Fairview in 1984.

The reception areas were in the northern sectors of the city in the newly established Chinese, Indian, coloured and African zones. One of the major features of all these areas, with the exception of the Chinese group area, was the complete control exercised by the municipality, or in the case of the African areas the central government, over development. Land for housing was owned by the authorities, which planned layouts and in most cases built the houses. It should be noted that the Port Elizabeth municipality, through its acquisition of the Bethelsdorp commonage and several other large tracts on the north-western side of the city, possessed sufficient land to ensure tight control over the housing programme. Further purchases in the 1970s ensured the maintenance of this control into the 1990s. Thus no townships were established on privately owned land and the sole private initiative was the zoning of limited areas for individual building. In the African areas the lack of rights to private land ownership restricted individual initiative with only one street of privately built houses in the entire set of suburbs until the 1980s. The contrast with the continuing suburban and smallholding sprawl of the white and Chinese areas, based on a free land market, has been most marked.

The state housing programme has been extensive with a massive phase of construction in the 1950s and early 1960s in the African areas when half (14,000) of the present stock of formal houses was built. In the 1960s and 1970s the coloured and Indian areas were developed on similar lines. The programmes in the African areas were such that the squatter camps which had been erected in the 1940s were reduced in size, although not eliminated. Until the mid-1970s, the housing shortage was artificially kept under some degree of control through the policing of influx control measures to prevent rural-urban migration and the overcrowding of existing housing.

Apartheid at its zenith

The degree to which the Union period philosophy of segregationism and the later philosophy of apartheid was enforced in Port Elizabeth can be gauged from the inter-group indices of dissimilarity. The index of dissimilarity refined by Duncan and Duncan (1955) provides a measure of the dissimilarity between the spatial distribution of different groups within an urban population. It is calculated on a scale from 0.0 (identical distribution, no segregation) to 100.0 (complete separation, total segregation). The indices for Port Elizabeth show a steady rise in segregation values throughout the present century (Table 3.2). Even in 1911, however, the white and African populations were markedly segregated from one another. The impact of the 1950s is particularly noticeable in the obvious increase in dissimilarity values. Thus indices by the 1960 census already indicated a highly segregated city.

Although the government never succeeded in its attempts to obtain total segregation, by the time of the 1985 census only 3.8 per cent of the entire population of Port Elizabeth lived outside its designated areas (Table 3.3). If provision is made for African underenumeration, the proportion was probably substantially lower. Thus to a large extent the apartheid model had been applied to the city. The anomalies were mainly confined to some 12,000 people living in the white group areas. The majority were African domestic servants and those resident in major institutions such as hospitals and prisons sited in the white areas. In other sectors, problems of definition within families notably through mixed marriages, particularly between the coloured and Indian groups, resulted in some overlap in the two group areas. The number of residential survivors in the white areas was comparatively small, while the numbers of residential permits for coloured and Asian people to live in the white areas was also small. The exception was the Chinese population largely concentrated (888 persons) in the Chinese group area at Kabega, which had been deproclaimed in 1984 following the redesignation of that community as white for purposes of the Group Areas Act. Even in the squatter camps which had developed around the city since the mid-1970s the division of functions between the Port Elizabeth and Ibhayi municipalities ensured the establishment of largely segregated African and coloured sections to the camps. In 1985 only the northern section of the Kleinschool camp indicated any degree of integration between coloured (2,510) and African (1,254) squatters. This has subsequently been altered with the

Table 3.2   Inter-ethnic indices of dissimilarity in Port Elizabeth 1911−85

Index	1911	1921	1936	1951	1960	1970	1980	1985
White-African	79.5	81.3	83.2	85.1	93.0	95.4	96.8	97.2
White-coloured	51.7	54.5	69.1	77.1	87.3	94.3	97.7	97.6
White-Asian	[1]	58.5	67.5	72.2	81.7	84.7	94.9	97.8
Coloured-African	68.8	68.6	61.3	72.8	88.0	93.1	96.2	97.0
Coloured-Asian	[1]	39.1	47.3	50.4	61.5	77.1	82.3	79.6
Asian-African	[1]	80.6	79.8	80.5	93.4	96.7	98.1	99.1

Notes
[1] Asians included with coloureds in 1911 census. Asians include Chinese and Indians.

Table 3.3   Area of residence in Port Elizabeth, 1985

Population group	Area of residence				Total
	White group area	Coloured group area	Indian group area	Black development area[1]	
White	130,018	347	20	395	130,780
Coloured	3,435	124,915	413	3,238	132,001
Asian	1,361	536	4,804	185	6,886
African	7,151	1,950	125	222,911	232,137
Total	141,965	127,748	5,362	226,729	501,804

Notes
Figures based on census enumeration tract data. Chinese were enumerated as Asians, although the Chinese group area had been reproclaimed as white.
[1] African residential areas (development areas) are not proclaimed under the Group Areas Act, but under the Black Communities Development Act of 1984.

proclamation of a coloured group area and the removal of many of the African inhabitants.

The allocation of land to the various groups is one of the key factors in the town. At the beginning of 1990 some 27,760 hectares in Port Elizabeth has been zoned for residential use on a racial basis (Table 3.4). The apportionment is highly unequal with a remarkably poor provision for the African population. The white population with over 12,000 hectares, mostly promulgated in 1961, has from the inception of the zoning programme enjoyed the most favourable conditions, with only 9−12 white persons per proclaimed hectare. The African population, despite substantial extensions to the proclaimed area, has recorded densities approximately 10 times as high.

Port Elizabeth in 1990

The population of Port Elizabeth in 1990 is a matter of some conjecture resulting from consistent underenumeration in the African areas. Growth since

Table 3.4   Group and Development Areas, Port Elizabeth 1961–90

Group	Area proclaimed 1990	Estimated population 1990	Persons/ hectare 1961 (census)	Persons/ hectare 1990 (estimate)
White	12,413	145,000	9	12
Coloured	10,115	150,000	19	15
Indian	377	7,000	23	19
Chinese	0	1,000	4	[1]
African	4,855	500,000	93	103

Notes
[1] Chinese group area zoned white by 1990.
African total median of range of estimated numbers.

1985 has been consequent upon natural increase as well as a substantial in-migration of approximately 20,000–30,000 Africans per annum following the lifting of influx control measures. Thus estimates of between 450,000 and 640,000 African inhabitants in 1990 suggest a substantial undercount in 1985 as well as serious deficiencies in basic information. The problem of overcrowding becomes the more evident when it is remembered that the majority of the African population continues to live in New Brighton and vicinity, at a gross density of 200 persons per hectare, while the privately developed, middle-income, suburbs of KwaMagxaki and KwaDwesi registered only 50 persons per gross hectare. The squatter areas record densities of 400–700 persons per hectare. Even if the population ceased to grow, the reduction of densities through the elimination of squatting and overcrowding would necessitate substantial spatial expansion of the African areas.

The problem of squatting has been addressed through the provision of basic site-and-service plots. Squatters sharing communal facilities now number approximately 250,000. Accurate estimates of the number of squatters are remarkably difficult to obtain as the majority of figures are based on shack counts and the estimates of the numbers of persons per shack. The counts are problematical as several shacks are often under one roof and the estimates of between four and eight persons per shack lead to widely differing results. Alarmingly, it would appear that approximately half the African population of Port Elizabeth lives in shacks. The largest concentration is at Soweto-on-Sea with a population variously estimated at 80,000 to 100,000. Appalling conditions prevail in the shack areas at present, with minimal drainage and sewage services and only one tap per 600 persons (Riordan, 1988b). Some 25,000–30,000 people reside in the Motherwell Emergency Camp established in 1982. Smaller numbers (approximately 6,000) of coloured squatters reside on the fringe of the coloured suburbs. Approximately 5,000 African and coloured people live in the bushland on the margins of the built-up area. The number of backyard squatters has been a major question for many years. Even in Motherwell many temporary shacks have been retained after the construction of more formal housing, effectively doubling the density of such areas. The prevalence of backyard rooms throughout the African areas is particularly noticeable in Walmer Township where there are more than four times as many backyard shacks as formal houses. Recent official attention has been directed towards the upgrading of squatter areas rather than their demolition. Shacks thus have a permanent and growing place in the housing market.

Clearly it is Motherwell which will have to accommodate the growth in the African population in the foreseeable future. The planning of this new town within the conurbation has been difficult as various demands have been placed upon the local officials, town councils and government agencies. The establishment of the Emergency Camp of shack dwellings created subsequent problems. The initial 1,400-site camp has grown to over 4,000 sites, while some 16,000 formal sites have been serviced in the town itself. The establishment of Motherwell followed the official realization that the government would be unable to do more than provide site-and-service plots, rather than houses for the African urban population.

Motherwell was therefore planned for a range of income levels from those able to erect architect-designed houses to controlled site-and-service schemes. Approximately 5 per cent of plots were provided for the élite group, currently able to build houses valued at over R50,000, while another 10 per cent were designated for medium-income houses, with prices over R25,000. The remainder (85 per cent) were provided for the low-income population. It is the low-income sector which has a marked shortage of at least 8,000 houses. This figure may be nearer to 20,000 if the definition of controlled squatting is widened, while the restriction of one family per plot would result in the housing shortage assuming even greater proportions. The most recent plans for Motherwell continue the pattern established in the initial layout, namely the zoning of the small wealthier suburbs closest to the city centre, with poorer areas planned further and further from the centre and the main industrial areas. The latest housing schemes are now being laid out some 15–20 kilometres from the city centre. In the next 10 years some 50,000 additional serviced plots are planned on new tracts (6,700 hectares) to the north and west of the present neighbourhoods. A total population of just under 500,000 is therefore envisaged at Motherwell. It might be pointed out that if the higher estimates of squatter populations are accepted, Motherwell could do little more than accommodate the existing housing backlog, estimated at over 50,000 units in the African areas. In the coloured areas a 1990 shortage of 7,000 houses can be met from within the lands available, while the small backlog in the Indian area will be satisfied within a few years.

It is here that the situation of Port Elizabeth with regard to the coastline has become of significance. The city, as a consequence of its coastal alignment, has remarkably limited opportunities for expansion. Algoa Bay on the east and the belt of sand dunes on the southern coast have effectively limited growth to a sector of ninety degrees from the centre. Thus the various sectors defined under the apartheid city model consist of only a few degrees each. The white areas, as a result of limited population growth and substantial infilling, have advanced little beyond the 15 kilometres reached in the 1960s. However, the comparatively narrow sectors for the coloured and African populations have advanced over 20 kilometres from the city centre.

Furthermore, the problems which Port Elizabeth has experienced with regard to its situation within the national economy have resulted in structural changes within the last 20 years. The industrial boom associated with the establishment of the motor car industry since the 1920s faded as the industry changed from an assembly of imported parts to local manufacture, close to the markets in the Transvaal in particular. Although new industries have taken up the losses the rate of growth has been slower than the national average. Accordingly the city has not experienced the economic growth associated with certain other parts of the country. Furthermore, the distance between the city and Ciskei and Transkei has been such that long-distance migrants have tended to seek more prosperous destinations. The immediate hinterland of the city is comparatively poor, based on extensive pastoral farming, resulting in few short-distance migrants in the last 20 years, despite the shedding of farm labour by white farmers in the region. Estimates of in-migration are remarkably variable but suggest at least 20,000–30,000 Africans per annum requiring housing, in addition to the natural increase of the resident population of approximately 20,000 per annum. It is noticeable that the coloured rural population of the eastern Cape has tended to migrate to other centres, notably Cape Town.

Industrial growth, which was an essential part of the early development of the city, has not kept pace with population growth in the last 20 years. Unemployment among African adult males is currently estimated at approximately 50–60 per cent. Whereas until the 1970s there was an attempt at the combined planning of industrial and publicly owned residential areas, thereafter industrial development has lagged. Distances to work have consequently increased for the coloured and African workers housed on the fringes of their respective suburbs. The short-distance commuting associated with the pre-1970s city is therefore giving way to longer journey times for an increasing proportion of the black population. Thus people living in the most recently developed residential areas experience journeys to the city centre of over 90 minutes, compared with under 30 minutes for the residents of the older parts of New Brighton. The burden of lengthening journeys to work falls upon a broad spectrum of the population as no sectoral social zoning is evident in town planning. Middle-income suburbs have been built adjacent to shack areas, most notably KwaMagxaki and Soweto-on-Sea. Owing to their greater accessibility, it is scarcely surprising that it is the older, and more central, African suburbs where the greatest pressures from squatters are experienced.

Transport is of vital importance to the functioning of the city. Subsidizing public transport to compensate for the peripheral position of the poorer sections of the community was accepted by the authorities in the city as early as 1896, when tram subsidies for one of the new locations were agreed upon. Subsidized train and later bus transport was meant to ease the financial burdens, if not the time costs, for people located for ideological reasons away from the centre of the city. In the 1950s and 1960s the coloured and African suburbs were planned with railway reserves, but the costs of construction precluded the provision of railways. The bus system thus assumed the burden of mass transport. This has now been largely superseded by the privately owned African mini-bus taxi service which has offered greater flexibility and convenience to passengers, often halving journey times. A new major bridge across the Swartkops River between Motherwell and New Brighton, completed in 1990, has improved the former's accessibility to the remainder of the urban area. A light rail system to link Motherwell with the city centre has been approved, subject to funding, in order to transport those on the new periphery to the centre and the main industrial zones.

Within the last few years a small number of black persons have moved into flats in the vicinity of the Central Business District. In addition there has been some black migration to the wealthier white suburbs including Summerstrand. However, there has been no significant development of a 'grey area' as distinguished in other metropolitan regions as the movement has numbered a few hundred rather than thousands. Remarkably, none of the Group Area Board investigations into contraventions of the Act since 1988 has led to evictions or prosecution.

Future directions

The future of Port Elizabeth is not easy to predict and such predictions as now appear may be far wide of reality. Growth is likely to remain below the national average both in industrial and population terms. It may be assumed that no major national development programme will be directed towards the continued ethnically differential growth of the city. It must also be assumed that the legislation associated with apartheid will be repealed and that a free market in residential land will be maintained and extended to all members of the population. In terms of local government, the fragmentation on ethnic lines, together with the complex pattern of African local authorities has made the overall planning of the area fraught with political problems. Conflict between the Cape Provincial Administration and the Ibhayi, Motherwell and Port Elizabeth municipalities over the acquisition of land for future expansion and the transfer of population illustrate the problems of systematic planning. The Algoa Regional Services Council has begun the process of upgrading the infrastructure and amenities of the black areas, but its limited financial base suggests that the establishment of one overall metropolitan authority will be necessary if the process is to be accelerated and comprehensive plans developed for the future.

In terms of population the present estimated total of approximately 800,000 inhabitants will increase to between 1.1 and 1.4 million by the year 2000, depending upon the rate of in-migration. As the majority of the population will be poor, without employment in the formal sector, the zoning of new sectors for informal housing will be of the utmost importance. There are two possibilities available to satisfy this need. The first involves pursuing the existing plans for the extension of Motherwell as a basic self-help housing area, with provision for upgrading as individual incomes allow. The area available to the South African Housing Trust is sufficient to cater for the immediate envisaged needs, but necessitates the extension of the city more than 25 kilometres from the Central Business District, with all the concomitant problems involved in the provision of mass transportation for people with low incomes.

The second possibility involves the utilization of the large tracts of waste land owned by the local and central authorities, including group area buffer strips, which could be made available for housing. This possibility has the advantage of providing land with proximity to existing services, the city centre and industrial areas. It would, however, be likely to meet resistance from householders in adjacent suburbs anxious to preserve their environment and property values. The major blocks of state land are to the south of the city adjacent to the airport and Walmer Township, which is experiencing its first physical expansion in 40 years at the present time. At the gross densities envisaged in Motherwell, these lands, covering over 6,000 hectares within eight kilometres of the city centre, could make a substantial contribution to the provision of housing land, accommodating the entire growth of the city in the next 10 years. Technical problems of building on sand have been overcome elsewhere and are not insurmountable. The buffer strips, particularly the extensive zones between the white and coloured group areas, similarly offer considerable tracts for future use. Here approximately 4,000 hectares are available within 12 kilometres of the city centre, part of which is the remaining section of the colonial endowment to the Church of England and the rest municipal land. Although some of the land is too steep for house building over 3,000 hectares is suitable, again providing for a major expansion of the built-up area, but no outward extension. Finally, smaller buffer strips often of several tens of hectares are vacant within 10 kilometres of the city centre, and are capable of development for housing or for recreational purposes.

The opening of all suburbs to all ethnic groups appears to present fewer immediate planning problems. In the first place the vast majority of the population will be unable to afford to move. However, movement between the non-African areas together with a more limited movement of wealthier Africans is possible. Studies in the United States suggest that even after the removal of legalized discrimination, integration is a slow and limited process (Massey and Denton, 1987; Darden, 1989). Such integration as has taken place so far in Port Elizabeth suggests three zones of activity. First, the central city flatlands (where the demand by single or newly married persons for accommodation can be met), may experience a greater inflow from the black suburbs, where such accommodation does not exist, except in the migrant workers' hostels. Secondly, upper middle-class suburbs, including Walmer, Mill Park and Summerstrand, with prestigious schools, will present environments where black executives and key workers can be accommodated and where no equivalent is present in the black suburbs. Thirdly, working-class suburbs may experience change where the white inhabitants may reproduce the classic process of invasion and succession as already evident in the politicized dispute over the empty, but white, state-owned flats in Algoa Park, adjacent to the crowded coloured suburbs.

The Port Elizabeth City Council, like that of Cape Town, opposes the creation of localized free settlement areas, believing that all residential areas should be open. Thus no official moves have so far resulted to establish free settlement areas. However, Fairview, demolished in the 1970s and 1980s as part of group area clearances, remains the subject of investigation. There appears to be little doubt that it will eventually be redeveloped as a multi-ethnic area. The tract of 600 hectares is relatively small, but may be compared with the combined area of KwaDwesi and KwaMagxaki, now housing some 24,000 people in a middle-class suburb. In a special emotive position is South End, where a return movement is possible. The mosques belonging to the Cape Malay community have been carefully maintained and are still in use, and may provide the focus for such a return migration. Just as it was the present white suburbs from which the black population was evicted as a result of the im-plementation of apartheid policies, so it seems reasonable to assume that it will be the underutilized white suburbs that will have to accommodate some of the population pressures from the overcrowded black suburbs. This need not result in radical change in the form of the present white suburbs, if open areas are used for the construction of houses and flats similar to those adjacent to them.

Port Elizabeth has substantial reserves of land for new housing, despite the problems of its coastal situation. The choices as to which lands to develop for middle-class housing will probably not be contentious as the present white, Indian and coloured areas all have suitable lands available. The significant feature will be the degree of movement between the presently defined group areas. Political controversy is most likely over the siting of low-income schemes, more particularly the expansion of controlled squatting. Extension to the south would clearly break down the ethnic sectoral pattern of the apartheid city, in favour of a series of concentric zones based on housing quality. The outlines of post-apartheid Port Elizabeth could therefore begin to emerge in a remarkably short time given the political and demographic pressures upon the city.

4
EAST LONDON
R. Fox, E. Nel and C. Reintges

The contemporary spatial form of East London and its 'bantustan' township appendage, Mdantsane, manifests the contradictions and dictates of apartheid urban planning and the attendant restructuring of labour relations. Racial residential segregation in East London has had a long heritage, originating in the colonial era and being refined and augmented by successive local and central authorities to produce the contemporary apartheid city form (Nel, 1990). East London's proximity to the Ciskei and Transkei bantustans (Figure 4.1) has been a factor exerting a significant impact upon the evolution of urban apartheid in the region.

Segregation in the pre-apartheid era (1847—1948)

The roots of residential segregation in East London, in common with many of the older cities in South Africa, originate in the colonial period (Davies, 1981; Christopher, 1987a). East London was established as a European settlement in 1847 in response to the need of the colonial authorities for a military post in this corner of the Empire, which was being subjected to tribal incursions from the Xhosa territories. This allowed for racial policies differing from those instituted by the authorities in Cape Town (Nel, 1989). From as early as 1849, under military authority, racial segregation was instituted in the town under the pretext of defence (Tankard, 1985). East London, which was part of the colony of British Kaffraria, was only incorporated into the Cape Colony in 1866. Under the colonial and military authorities, location regulations for Africans and a rudimentary system of influx control were instituted. This ensured that when the first municipal council assumed office in 1873 both segregation and discrimination were well established.

Between 1873 and 1895, utilizing the enabling measures granted to the city by the Cape Colonial government, numerous segregated African locations

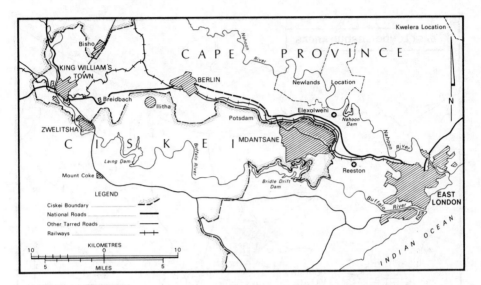

Figure 4.1   East London in regional perspective

were created by the municipality. Segregation of the Indian community in 1895 was authorized by the East London Municipal Amendment Act, a measure that was unique in the Cape Colony with the exception of Kimberley (Davenport, 1971). Despite the ultimate failure of this attempted segregation of Indians, white racialism and resentment of Indian commercial competition ensured that informal segregation tended to confine Indians to the racially mixed area of North End (Figure 4.2). Civic action was motivated by fears that the city would emulate Durban, which they regarded as a 'town in the python coils of an Asiatic menace' (*Daily Dispatch*, 16 August 1928). Prior to the apartheid era no formal segregation measures were applied to the coloured community. The initiation of the Parkside municipal estate for the exclusive occupation of this ethnic group, helped to ensure their partial and informal separation from the broader community. Prevailing hostility and discriminatory market forces obliged the remaining coloureds to live in what were often dilapidated dwellings in North End and the African locations (Bettison, 1950).

If the plight of the Indian and coloured communities was deplorable, that of the Africans was far worse. Throughout most of the present century Africans have been obliged to reside in three African locations, namely the East Bank (later known as Duncan Village), West Bank and Cambridge. These locations predate the urban segregation of Africans which became obligatory on a national level in 1923 in terms of the Natives (Urban Areas) Act. Since the Council did not adopt the 'Durban system' of a beer monopoly to subsidize council housing, most Africans were forced to live in owner-built shacks on land rented from the council. Conditions in the locations were appalling. For example, in 1916, the 9,500 residents of the East Bank location had access to only 11 water-points (Watts and Agar-Hamilton, 1970). Disease and destitution

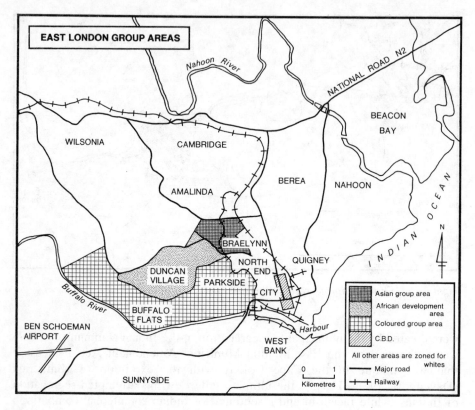

Figure 4.2   Group areas in East London

were rife under these conditions; the infant mortality rate stood at 45 per cent
of live births among the African community in the 1930s (Minkley, 1985). By
1948 the city had the highest rate of tuberculosis in the country. Popular
opposition to the status quo tended to be led by trade union movements
(Minkley, 1985) with resentment frequently being expressed in the local press
(*Izwe Lambe Afrika*, 1931; *Daily Dispatch*, 1931–7).

The need to rectify the prevailing conditions in the African locations caused
the government to institute two commissions. Their briefs were to investigate
the situation which prevailed and to make recommendations. They were the
1937 Thornton commission and the 1949 Welsh commission. The latter attributed the 'present repellent state of the locations (to) ... over half a century
of apathy on the part of the European ratepayers' (East London Municipality,
1949–50, p. 4). Both commissions urged immediate action to rectify the gross
overcrowding, including the upgrading of facilities, the extension of Duncan
Village into the Amalinda area (Figure 4.2) and the possible creation of a new
satellite location. This latter measure provided the rationale for the future
development of Mdantsane, where the bulk of East London's African labour
force now live (Gordon, 1978). These steps were delayed by World War Two

and the council's apparent financial straits, obliging the city, after 1948, tacitly to accept the urban redesign the apartheid planners deemed desirable.

East London in the apartheid era (1948–89)

The period after 1948 was marked by several clear tendencies, namely: the vacillation of the local authority, increasing popular dissatisfaction and the determination of the central state to implement apartheid. The two primary features of the era were, first, the attempt to apply the Group Areas Act to the city with the attendant relocation of disqualified people, and secondly the decision to develop Mdantsane as the future residence of all the city's Africans in what was to become the 'independent' bantustan of Ciskei. It is significant to note that segregation prior to 1948 had been the result and manifestation of local initiative and discrimination. Thereafter, although the local authority welcomed the state's commitment to rehouse thousands of destitute people in Mdantsane, it was fundamentally opposed to the Group Areas Act.

In common with all South African cities, East London merited the attention of the Land Tenure Advisory Board, later known as the Group Areas Board, which made recommendations regarding the future application of the Group Areas Act. The government decided to displace the entire African populace as a preliminary measure and so the application of the Group Areas Act came much later than in other centres. The primary reason for this was the delay in the removal of Africans to the new bantustan location. The areas the Africans vacated were zoned for Indian, coloured and (initially) Chinese group areas. One area was, however, zoned for whites in 1952. Following the 1952 location riots and the opposition of local white residents to the proposed location extension into Amalinda, Dr H. F. Verwoerd and the Group Areas Board agreed to zone Amalinda as a white group area. This step terminated the council's only existing location improvement plan, further aggravating the plight of thousands of Africans. The council's approach to the Group Areas Act appeared to accord with those of other 'liberal' English municipalities in the country, namely that: 'the council is of the opinion that zoning proposals for the city of East London are not called for, in view of the fact that there is no real problem in the city regarding racial zoning' (East London Municipality, Public Works Committee Minutes, 1961).

In order to enforce the Group Areas Act and to eradicate the racially mixed slum area of North End (Figure 4.2), a special state committee to liaise with the council and to purchase and redevelop North End was appointed. According to a municipal survey, 3,226 of the 3,816 Africans in the city's white areas lived in North End, making it a focal point for government attention. As land became available in the cleared African locations, the residents of North End were moved to newly established segregated housing estates and their former homes demolished by the government.

The most dramatic spatial and social change wrought by apartheid was occasioned by the decision to place the African populace in the bantustan township of Mdantsane (Figure 4.1). The aims of urban apartheid in East London were coupled with the government's bantustan strategy. The residents of Mdantsane became 'frontier commuters', working in East London, yet living in an 'independent' country. The failure of the council to improve the lot of the Africans, largely through financial incapacity, led to direct government intervention, designed not only to rectify the situation but more specifically to fulfil Verwoerd's ideal that: 'the new township will not be an urban Bantu residential area in a white area but will be developed as a Bantu township in a Bantu homeland' (*Daily Dispatch*, 9 February 1962). This step was in line with government policy on 'self-development' enshrined in the 1959 Bantu Self-Government Act, which provided the mechanism to implement the homelands policy. Construction of Mdantsane started in 1963 and has continued unabated to the present, allowing for the relocation of Africans from the city between 1964 and 1983. The removal of the African populace was never fully achieved. The magnitude of the task was beyond the grandiose ideals of the apartheid planners. Furthermore, the removals were resisted and eventually halted following mounting popular hostility and the refusal of Ciskei to accommodate any more people removed from East London (*Daily Dispatch*, 24 June 1983).

By 1987, 26,101 houses had been built in Mdantsane occupied by at least 110,000 displaced East Londoners (Walt, 1982). The township's population has exceeded that of East London since 1973, becoming, *de jure*, not only the largest bantustan township, but also the second largest formal African township in South Africa (Walt, 1982). Its present population of more than 250,000 exceeds that of East London itself. Apartheid has not solved the African 'problem' which it had set out to do. Tens of thousands of Africans still reside in East London, living under conditions little changed from those of 50 years ago because of the freeze on development during the removal years. Apartheid has thus had very distinct and pervasive effects on the social and spatial fabric of the metropolitan area.

The present grossly unequal distribution of land is a consequence of the policies described above. Table 4.1 shows that whites were allocated over three-quarters of the land although they only constituted one-third of the

Table 4.1   Ethnic groups and land allocation, 1988

Ethnic group	Population	Percentage of total population	Land (ha)	Per cent of total land	Per cent of land developed
African	100,000[1]	52.7	324.30	9.1	100.0
White	62,840	33.1	2,756.80	77.7	84.0
Coloured	24,210	12.8	336.95	9.5	88.0
Indian	2,900	1.5	131.70	3.7	46.0
Total	189,950	100.1	3,549.75	100.0	

Notes
[1] An estimate that is midway between the lowest figure (60,000) and the highest figure (140,000), see Reintges (1989).
(*Sources*: Reintges, 1989; Theart, 1990.)

population. By contrast, the African population occupy less than 10 per cent of the land even though they constituted over half of the population. The Indian group also benefited, with 1.5 per cent of the population on 3.7 per cent of the land, whilst the coloured group were the other losers, with 12.8 per cent of the population on 9.5 per cent of the land.

Not surprisingly, the population densities for the ethnic groups show that the white and Indian densities were low at $c.22$ persons per hectare, coloured densities were quite high at 71.9 persons per hectare and African densities were very high at 308.4 persons per hectare. These densities are likely to change only slowly in the short and medium term even as the structural constraints of the apartheid state are removed.

African housing

Housing and land policy in East London's African Township of Duncan Village are an apt example of the end product of apartheid planning and the institutional chaos in which policy is supposed to be enacted. Following the complete reprieve from removal to Mdantsane in 1985, upgrading proposals for the improvement of housing and services in Duncan Village were proposed. Initially the proposals were integral to the state's programme which was envisaged to calm troubled areas through the improvement of living environments. Since 1986 the proposals have been firmly couched within the orderly urbanization reform package (South Africa, 1986). While the details of the upgrading proposals (Reintges, 1989) are of little relevance here, the number affected, the basic parameters and the implications of undertaking the upgrading are of consequence.

The upgraded Duncan Village, according to the plans, was to accommodate 23,000 people. The state argued that this would displace only 9,000 people who would be accommodated at a newly identified African residential area called Reeston (Figure 4.1). These official figures were crudely derived by simple arithmetic: 112,000 (the number of people in Duncan Village at start of removals) less 80,000 (the number of people removed to Mdantsane) = 32,000 (23,000 accommodated Duncan Village and 9,000 = overflow). The derivation of these figures illustrates the hangover of state policy regarding influx control and homeland development. There were no other people to count in Duncan Village because they were 'illegal' and therefore were living in the homelands. Table 4.2 indicates a range of possible figures which more closely resemble reality. In June 1989 there were an estimated 15,000 shacks in Duncan Village concentrated in the section that, according to the plans, is to be demolished in its entirety (Reintges, 1989). Personal observation over the last year has revealed that this number of shacks has increased.

Upgrading within orderly urbanization is undertaken on a cost recovery basis and much of the development has been handed to private contractors as part of the state's recent privatization initiatives. These elements of the upgrading programme have the effect, similar to gentrification, of creating a middle-class environment which effectively evicts those who cannot pay for the new goods

and services provided. In effect land and services have undergone commodification and their exchange on the market has little regard for the social
consequences of the transactions (Swilling, 1990). Referring to Table 4.2 we
can see, using 1989 figures, that between 12,739 and 117,00 people will constitute 'overflow' on completion of the upgrading programme. These people must
be accommodated elsewhere but due to underplanning there is insufficient land
at Reeston which is the only area identified for African residential development. Secondly the Mayor of the Town Council has said that no site-and-
service area will be provided at Reeston. The result of the state's privatization
drive, bolstered by the Urban Foundation's recent loan initiatives (Swilling,
1990), is that the poorest people are not catered for since they will be forced to
move as a result of upgrading.

It is perhaps fortunate for these people that the state is not using the rapid
means provided by the 1989 Prevention of Illegal Squatting Amendment Act to
identify and proclaim the land for African residential purposes. They are using
the old slow bureaucratic route which leaves the land as yet unsurveyed. So
long as Reeston remains undeveloped, upgrading cannot take place as the
people have nowhere to move to. The underserviced conditions in which
people in the shack sections are living, however, demand that affordable land
and housing be developed very rapidly. As usual it is the false scarcity of land
created by the Group Areas Act and existence of the Ciskei bantustan that
compounds this process and, in the final instance, ensures that Duncan Village
will never become a middle-class residential area.

These physical, technical and planning problems need to be addressed as a
matter of grave urgency. Equally critical are certain political issues. Swilling
(1990) stresses that the commodification of services, land and housing means
that transactions leading to the displacement of people are individualized
events. Resistance on such a basis is difficult if not impossible. This is an issue
that must be addressed by the Mass Democratic Movement and the African
National Congress (ANC), if tens of thousands of people are to resist the

Table 4.2   Potential densities (persons/site) at Reeston

Estimated 'overflow'	4000 sites	5000 sites
12,739[1]	3.2	2.5
37,000[2]	9.25	7.4
67,000[3]	16.75	13.4
117,000[4]	29.25	23.4

Notes
[1] The Minister of Constitutional Development quoted the population figure of 35,739 ($-23,000 =$ 12,739).
[2] Setplan calculated a population figure of 60,000 ($-23,000 = 37,000$).
[3] The lowest estimate provided by the Gompo Mayor and the Gompo Chief Engineer is 90,000 ($-23,000 = 67,000$)
[4] The Gompo Chief Engineer's highest estimate is 140,000 ($-23,000 = 117,000$).
(*Source*: Reintges, 1989).

process and secure themselves access to affordable housing. Another political problem is the growing disjuncture between central state and local reality. The National Party and ANC politicians are raising expectations at grass roots level. Unfortunately, the nature of reform being pursued is largely irrelevant to the day-to-day reality of local development and at times, as in the Elexolweni case (Grahamstown Rural Committee, 1988–90) the diehard Verwoerdian bureaucrats are subverting well intended reform.

At Elexolweni the establishment of a semi-rural settlement has been instigated from the grass roots level by an African community. This type of settlement gives its inhabitants a number of survival options: subsistence agriculture, small-scale market gardening, work in the industrial areas of East London or Mdantsane. Unfortunately, the development is within white South Africa and so lies beyond the conceptualization capacity of local state officials. The officials have then used apartheid legislation to frustrate the development of what is, in fact, a type of settlement commonly found around many tropical African cities such as Nairobi. This type of problem is exacerbated by a lack of policy direction which could perhaps re-direct bureaucracies.

It seems to be the case that there are marked contradictions between public statements of policy and reform by politicians and the legislation, both recent piecemeal reform and older apartheid strictures, that are being implemented by the executives at local level. Much of the planning bureaucracy in East London and the eastern Cape is embedded in the old system and is slowly coming to terms with the new legislation. At the same time national negotiations are suggesting an agenda that will outstrip both the old legislation and its recent modifications.

Political and administrative framework

One of the basic contradictions which the South African state failed to resolve in the 1980s was that cities and metropolitan regions are functional, economic wholes that have been administratively and politically unpacked but which need to be repacked so that planning and economic development can proceed with some semblance of efficiency. Within the city itself it has been state policy to fragment the urban area into racially homogenous zones, each one of which can be differentiated internally along socio-economic lines (Cloete, 1986). Each group area was to be managed separately but control was to remain in white hands.

Thus in East London, at the local government level, the white municipality manages its own affairs and also the interests of the coloured and Indian groups through the management committees (Table 4.3). By 1990 there was no effective planning department within Gompo Town Council and so the little planning that was being undertaken came through the Cape Provincial Administration in Port Elizabeth (Els, 1990), a body that was not constituted for this task. As part of the incremental reform process Regional Services Councils have been set up in the eastern Cape in order to redistribute resources for infrastructural upgrading to the most disadvantaged local authorities. The

Table 4.3   Political and administrative structures

SOUTH AFRICA	CISKEI	TRANSKEI

Parliamentary Representation
White Constituencies:
Albany
East London
King William's Town

Transkei and Ciskei
Under Military Councils

Coloured Constituency:
Grens

Indian Constituency:
Malabar

(No African constituency in South Africa since Africans are excluded from the Tricameral Parliament)

Regional Development
Regional Development Advisory Committee – Region D

RDAC-D	RDAC-D	RDAC-D
(South Africa)[1]	(Ciskei)	(Transkei)
9 Work	9 Work	9 Work
Committees:	Committees:	Committees:
Housing etc.	Housing etc.	Housing etc.

9 Joint Work Committees

1 Regional Liaison Committee

[1] (No meaningful African representation in RDAC-D, South Africa.)

Local Government

White Municipality:	Municipal Councils:	Municipality:
East London	Zwelitsha	Butterworth
King William's Town		

Coloured Management
Committees under
White Municipal
Control

Under Central State
Control:
Mdantsane

Indian Management
Committees under
White Municipal
Control

African Town Council:
Gompo, governing
Duncan Village

councils, which exclude the bantustans, have not been in operation long enough to assess their success but it seems likely that the addition of a co-ordinating umbrella organization like this will be ineffective given the already existing complexities and contradictions at local and regional level.

In addition to the urban authorities in South Africa are the local government structures in Ciskei and Transkei. The largest residential area in the greater East London metropolitan area lying within Ciskei is Mdantsane. Its residents have had no officially recognized local representation since they have a history of resistance to the Sebe regime; their planning was therefore undertaken by a central state organization, the Directorate of Planning, in Bisho. It appears that another state department, the Directorate of Interior Affairs and Land Tenure, could easily act to block development, especially of low-cost housing, through manipulating the orderly release of state land (Els, 1990). This was a problem experienced by both the Urban Foundation and private entrepreneurs who depended on the speedy and reliable release of land to make low-cost housing financially viable. The other main towns in Ciskei and Transkei in Region D, Zwelitsha and Butterworth, both had councils and an executive that were under control of the respective planning ministries.

At the regional level the fragmenting impact of apartheid is particularly pronounced. In Ciskei the major residential areas of Zwelitsha and Mdantsane were both created as acts of official policy: Zwelitsha founded as early as 1946 under the border industry policy, Mdantsane from 1981 as part of the bargain struck with former President Sebe to create the independent Republic of Ciskei. Industrial development points have also been encouraged since 1984 under Pretoria's regional development strategy for southern Africa (South Africa, 1984a). Through this policy, industry has been encouraged to locate at different places in each of the three different 'countries' in Region D: Butterworth in Transkei, Mdantsane, Dimbaza, and Berlin South in Ciskei, King William's Town, East London, and Berlin in South Africa (SECOSAF, 1985). Since the region is an economic whole split into three supposedly independent states the regional development advisory committee (RDAC) structure has been used in an attempt to co-ordinate development (Table 4.3).

The South African government (1988b) sees the role of RDACs as follows:

1. To collect and compile knowledge of the region.
2. To co-ordinate and help formulate regional development programmes.
3. To identify development problems.
4. To supply development advice to concerned communities and organizations.
5. To foster regional awareness and identity (i.e. having ensured that there is no regional identity the RDAC must try and create one).
6. To publicize development potential.
7. To co-ordinate policy across state boundaries.
8. To serve as a channel of communication between the authorities within the region (i.e. having split up the region, there is a need to put it back together again).

In practice, both critics of the system (Black *et al.*, 1986), and its participants (Theart, 1990), have found it impossible to implement. The following quotation from the Draft Guide Plan for the Greater East London Metropolitan Area gives a clear idea of the strategy's stultifying impact on development:

> The cumbersome regional structure ... has resulted in an administrative system which includes several hundred individuals in the activities of 27 working committees, 3 Regional Advisory Committees, 9 joint Working Committees, a regional liaison committee, several *ad hoc* Task Teams and working groups, and the East Cape Strategic Team. It is quite evident that the system itself has prevented its participants from addressing ... the critical socio-economic and political issues which impact the region (Theart, 1990, section 5.2.1.1(g)).

Finally, Table 4.3 shows the parliamentary representation in the region. This is also split into three, one system for each country, with South Africa being further split into three with different constituencies and MPs for the different race groups in the tricameral parliament (Fox, 1990). The constituencies themselves are of vastly different size since they reflect the different population densities of the ethnic groups. In the eastern Cape region, which stretches from Port Elizabeth to East London, there is only one Indian constituency, Malabar, there are three coloured constituencies, Addo, Visrivier, Grens, and six white constituencies, Sundays River, Albany, Cradock, Queenstown, King William's Town, and East London. This is another instance of inconsistency in the apartheid system. The 1980 electoral commissioners maintained that a regional identity and sense of community would be fostered where electoral divisions follow administrative boundaries (South Africa, 1980a). Unfortunately for them, the legislation they were enforcing meant that this could not happen to society as a whole. The three ethnic groups had to identify with different overlapping or cross-cutting areas.

The situation at present is highly unsatisfactory in that none of the region's African population have recourse to regular, democratic electoral procedures. Africans cannot participate in the tricameral parliament and both Transkei and Ciskei are governed by Military Councils, with apparent ANC sympathies, which came into power through coups d'etat in 1987 and 1990 respectively.

The future: uncertainties and constraints

It is difficult to conclude given the growing uncertainties that arise as political rhetoric and the intention to induce reform gain momentum at national level. One effect of the coup which toppled the repressive Sebe regime is that people who lived in churches in King William's Town for months in their desperation to escape repression are now willing to return to their land at Peelton in Ciskei. However, South Africa is in such a state of flux at present that it is difficult to speculate on whether this type of move is a precursor to others anticipating the re-incorporation of Ciskei into South Africa. It is possible, however, to anticipate some of the planning constraints likely to influence the development of the metropolitan area.

The greater East London metropolitan area has been split into a number of spatially discrete units: East London, King William's Town, Mdantsane, Zwelitsha. Future growth will be either within or between these units depending largely on whether Ciskei is re-incorporated within South Africa. The chief technical obstacle to urban growth if there is re-incorporation will be the different land tenure systems in Ciskei and South Africa.

In Ciskei there is very little freehold land in urban and peri-urban areas; most of the land is held under Government Notice R293. This gives the state total control over the purchase, sale and registration of this type of land (Els, 1990). In the rural areas land tenure for individuals is equally insecure with the final authority usually resting with the Agricultural Officer, magistrate, or Minister (Cokwana, 1988). By way of contrast, the land in the Border region comprises white agricultural land under individual freehold tenure, trust and mission land. As the ANC, Pan Africanist Congress (PAC) and the present government are fully aware, much research into the land reform process needs to be undertaken since there is a huge demand for access to land and shelter.

Whatever the process of land reform entails, it is likely that spontaneous development on East London's periphery will be prevalent. The type of development will probably be peri-urban with agriculture practised on small plots to augment income from urban employment. This is a pattern found elsewhere in Africa (Swindell, 1988) and which is already occurring at Newlands, Kwelera and Elexolweni today. There is of course the possibility that shacks will become the most lucrative crop, a prospect which alarms the government. Rural-urban migration will inevitably intensify as people leave the overpopulated and devastated resettlement areas of Ciskei.

Within the urban areas themselves the repeal of the Group Areas Acts would ease the housing problems for middle-class Africans and coloureds as they should be able to move into the equivalent white suburbs. There already exist racially mixed areas on the eastern side of the CBD, just beyond the tracks of the railway. The authorities have turned a blind eye to this development in spite of its illegality. Other middle-class areas would generally become more racially mixed; but high-income areas will probably remain largely white in the foreseeable future. The poorest income groups should also benefit in the medium term from the repeal of group areas legislation, since the cumbersome land proclamation and purchase system, which is currently holding up the development of Reeston, ought no longer to be a problem. Extensive low-income suburbs, predominantly African, will be a dominant feature of the future urban structure of East London. Some will be planned by the state, some peri-urban in nature, with squatting within and beyond the urban area.

The key to the future spatial organization of the East London metropolitan area lies with the re-incorporation of Ciskei into South Africa. In the past, the Ciskei border has acted as a fixing line, on either side of which different processes have been experienced under different political, economic and land ownership systems. The patchwork of urban development in the East London area is the result of this. It would be economically and administratively more efficient to have one system and it is imperative socially that the poor be given freedom of access to land and thereby shelter. Major reforms which involve

centralized state ownership of land and a cumbersome land allocation bureaucracy are not likely to satisfy this need, nor will a free-enterprise, private
ownership system since the poor cannot compete. Reform therefore needs to
set up an efficient re-allocative mechanism by which access to land is possible
for all. Needless to say, this will be unpopular amongst those groups who have
been manipulating the land market, for example the chiefs in Ciskei and the
state in South Africa, who can be expected to resist any reform.

Note

Archival material has not been cited; details may be obtained from the authors.
Principal sources include the Cape Archives Depot and archives of the East London
Municipality, including Mayor's Minutes and the minutes of various committees.

5

DURBAN

R. J. Davies

Home to more than three million people, the Natal coastal city of Durban is today South Africa's second urban place by population size. Proximity to the national economic heartland has meant also that it ranks after the Witwatersrand as the second most important centre of economic production and is the country's major port and holiday resort.

Though not the largest, Durban is arguably South Africa's most complex city. Here, forces of a mixed capitalist mode of production, rooted in a colonial past and supported by powerful political structures, have generated a social formation and a set of contingent geographical imperatives which have, over time, evolved the archetypal apartheid city. Durban, like any other individual place in South Africa, has been a receiver of and responder to forces generated in the broader society. Its own particular social formation, however, has been a major source of structural forces which have exerted a determining influence upon South African urban form and life. It is here too that manifestations of contemporary urban development trends are becoming strongly evident and are pointing to possible urban futures. These are themes which are explored in this chapter.

In the analysis which follows Durban is shown to be not only historically rooted in a colonial past but to be an outcome of a structural continuity of the colonial process in a context that might be described as 'internal colonialism' (Williams, 1983). Structuring of the apartheid city has indeed carried the process to an unprecedented extreme (Davies, 1986). It is a city which, since 1976, however, has been taken up in powerful forces of national structural conflict the outcomes of which have become increasingly evident. Hitherto, responses have taken the form of adaptive change, but since the start of 1990 they have begun to shift towards a phase of fundamental transformation.

In 1824, at a time when British colonial commercial capitalism was taking root in the Cape Colony, Durban was established as a privately promoted ivory trading post on the Natal coast. It grew as a colonial port and commercial centre through the late nineteenth century benefiting substantially from the establishment of the mining industry at Kimberley from 1870 and the Witwatersrand from 1886 and the development of a rail transport network. Its population increased from a few thousand in 1850 to 116,000 by 1911.

Since Union in 1910, Durban's economic base has undergone dramatic change and economic diversification. Secondary industry, a large-scale holiday and hotel industry, financial activity, a transport function and large-scale developments in the port have been added to the already important commercial function. The population of the greater metropolitan area had grown to an estimated 3.4 million people by 1989.

The social formation

In Durban's colonial setting and expanding commercial economic base lay the structures of a typical settler-colonial urban social formation. It contained determining geographical imperatives.

By applying a qualified franchise first at the level of local government (Ordinance 1 of 1854) and later, at the territorial level, in the Natal constitutions of 1856 and 1896, Africans were effectively excluded from the franchise. Colonial whites were assured of dominance in political power relations. Under these conditions Durban was incorporated, in 1854, as a municipality with 1,200 white and over 3,000 African inhabitants.

The primary resource upon which settler colonial production rested was land. In the Durban case the early traders, through manipulative relations established with the indigenous Zulu leadership, attained rights to land use at Durban and in its hinterland. The concept of 'use' was appropriate in traditional, redistributive, Zulu society. The settlers, however, interpreted the agreement in terms of rights to ownership. Parts of that land were later coercively appropriated by Boer trekkers colonizing Natal from 1838. Under British colonial rule from 1845 the principles of individual tenure were confirmed as land passed progressively into white hands. In Durban itself the incorporation of the borough in 1854 similarly vested control of the townlands in white hands.

To provide for the indigenous population, parts of the colony were, from 1846, proclaimed as 'native reserves'. The procedure served to concentrate surplus African populations and to define colonial space. The reserves too were intended to articulate future colonial labour needs with the traditional, redistributive economy in which labour could be stored and reproduced at little cost to the colony. Viewed purely as a potential labour force, the African population was not expected to gain access to or control over the essential means of production or to economic and political mobility in colonial space.

In this way the primary dimensions of settler colonial dominant-subordinate relations, in which class and race are closely associated, were set in place. Of course, not all white colonists were owners or controllers of means of production and many formed part of the worker class. By virtue of their membership of the dominant group, however, they inevitably gained a privileged socio-economic and political status.

Ongoing viability of the indigenous redistributive economy and a reluctance to enter labour meant that African entry to the colonial labour force was constrained. In Durban, the African population grew from 3,000 in 1854 to some 22,000 people by 1911 at which time whites outnumbered Africans in the city by two to one.

These circumstances were to lead to the importation of Indian indentured agricultural labour from 1860. As a rural, immigrant, poor working class population without effective political influence, Indians initially offered no material threat to the dominant colonial group. As their numbers increased and as they became urbanized from 1870 the position began to change, aided by voluntary immigration of 'passenger' Indians, mostly Moslem traders, from 1874. Between 1870 and 1911 the Indian population in the metropolitan area rose from 1,900 persons to over 54,000 persons and Indians became the largest single group, outnumbering whites by five to four. Their share of urban employment increased, trading activities expanded and property acquisition in Durban became evident.

The emergence of a competitive, ethnically defined fraction of capital, strong cultural differences and the poverty and poor living standards of many, raised white antipathies. An 'Asiatic menace' was identified (Swanson, 1983). In 1896 Indians were excluded from the colonial franchise. From 1897 the white municipal council of Durban through its licensing by-laws was able to constrain the expansion of Indian trade but not suppress it. Nor was the acquisition of residential property contained. Thus in law and practice the dominant white group succeeded in marginalizing, though not entirely subordinating, the Indian element in the colonial social formation.

Union of the four South African territories in 1910 represented a political accommodation that overcome a conflict threshold between two dominant groups in the colonial social formation − the English and the Afrikaners. It left the future relationships of the colonial subordinate and marginalized groups unresolved, and the essential political and economic relations of the colonial social formation were perpetuated.

It followed that the dominant group in post-colonial society would maintain coercive control over the social formation to conserve its material, political and social interests, including those of the privileged white worker class. A wide-ranging set of strategies and procedures of control was progressively imposed through legislation, regulation and practice as the scale and complexity of economic growth, urbanization and the diversification and specialization of urban society increased. Such controls in post-colonial urban society were only partially successful as the case history of Durban will show. By 1940 the outcomes of incomplete and permissive control were becoming strongly evident in the relations of the urban society and the structure of South African cities. Circumstances existed which in the white view justified the imposition of stricter control though ideas on how it might be imposed varied (Schoombee, 1985).

In the event, the concept of an apartheid society triumphed and became official policy from 1948. It was a natural expansion of earlier processes but was officially presented in purely racial terms as a means of avoiding and ameliorating inter-race/ethnic and cultural conflict (Kuper, Watts and Davies, 1958). In reality it was designed to overcome crises of control. In the urban context a comprehensive and interlocking framework of existing and new institutionalized political, economic, social and spatial controls, some with retrospective application that violated existing rights and freedoms, was imposed over time. It was intended clearly to define the parameters of the social formation, firmly to regulate its relations and to ensure that inherent geographical impera-tives were appropriately expressed and utilized in structuring and justifying the form of the cities and in important respects containing and regulating the lives of their subordinated and marginalized people. Indeed urban apartheid was to carry the neocolonial social formation and a structured neocolonial city to an extreme conclusion.

That such a system, apart from practical considerations, contained deep-seated inherent potential for resistance and conflict goes without saying. Those forces, taking many forms apart from overt violence, inexorably led to a new and more intense urban crisis. Thus from the mid-1970s the society has moved progressively into its current phase of transition. Implications for future urban structure are substantial.

Colonial Durban

Within the framework of the present metropolitan area, colonial Durban (the old borough) occupied an unusually small area controlled by a white municipality. Beyond its boundaries lay an unregulated urbanizing periphery. At greater distances, small suburban nuclei were sited in the corridors defined by the boundaries of the Inanda and Umlazi 'native reserves'. The land initially appropriated by whites was the basis of a property market in which Indians began to participate from 1870. It produced a land use structure characteristic of capitalist cities (Figure 5.1(a)).

An evolving and focal central business district (CBD), generated mainly by white capital, had by 1910 developed a dual structure as Indian traders, regu-lated from 1897 by licensing by-laws, became concentrated on less desirable land on its north-west edge (the Grey Street area).

Forces of social formation had generated a distinctive organization of residential space. Eighty per cent of whites lived in the old borough and, exercising choice consistent with dominance, commanded the most desirable, conveniently situated land. Economic differentiation took place by desirability of location. Indians from 1870, on the other hand, exercising *their* rights as a marginal group, had initially purchased property in the Indian commercial area. They lived in close association with work places creating high density living conditions which became a source of white affront. As numbers grew, however, they diffused outwards invading the neighbouring white residential areas of Greyville and Old Dutch Road. Though mixed, those areas housed 34

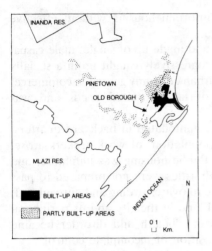

Figure 5.1a Durban: colonial city outlines, c.1900

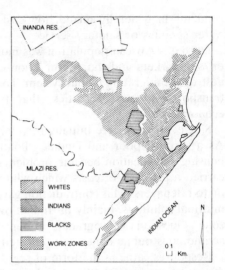

Figure 5.1b Post-colonial Durban, c.1950 – population groups

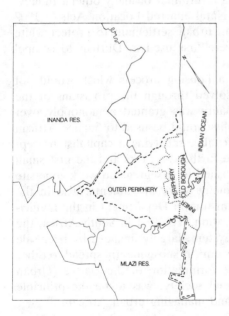

Figure 5.1c Post-colonial Durban – metropolitan structure

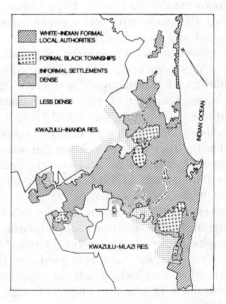

Figure 5.1d Durban – metropolitan administration

per cent of the metropolitan Indian population by 1911.

However, the greater number of Indians, 48 per cent of the 1911 total, for ease of entry had acquired or rented low priced and unregulated land in the peripheral zone. There they established themselves as market gardeners or residents in generally mixed, low-cost or informal housing without effective water supplies or sanitary services.

Durban's African population was mobile and made up of single, male casual or *togt* workers and short-term labourers progressively caught up in a socially destructive migratory labour system. To an urban economy geared to commerce, transport and port activities, that type of labour suited needs and was economically beneficial.

Single workers were initially privately accommodated in backyard quarters. As a surplus, poor and landless 'floating population' of work seekers arose, housing and location became problematic. Public housing was unfamiliar and carried a prospect of costs which white authorities were not prepared to pass on to ratepayers. This contradiction gave rise to overcrowding and unregulated, informal solutions, mainly on Indian-owned land in the city and its peripheral zone. Questions of vagrancy, nuisance, crime, health and disorder became acrimonious public issues and reflected a position of incomplete control.

Responses came in the form of registration regulations to control the inflow of *togt* labour in 1874 (the Togt Law of 1874) and a pass system (Acts of 1884 and 1888). In 1878 the first worker, rental barracks were erected in the dockside work zone. They were to be the forerunner of many others. Finally, following the lead of the Cape Colony, Natal enacted Location Acts in 1902 and 1905 to control and regulate African urban settlement to protect white social and health interests. The Acts were first used in Durban to compel Africans to take up residence in barracks.

The search for a self-financing African housing process which would not burden white ratepayers, was partly resolved through the provisions of the Native Beer Act of 1908. The municipality was granted a monopoly over brewing and sale of traditional beer. Profits would be used to finance African housing, neatly, though highly controversially, satisfying a capitalist precept that the user must pay. From this source further barracks and the first small family location of 36 cottages were established on a segregated work-zone site at Baumanville in 1914. These developments were instrumental also in the establishment of a separate Native Administration Department in the municipality in 1910 to regulate African urbanization. The first in South Africa, the administrative structure and its housing system, largely financed by beer sale profits, became known as the 'Durban System'. It subsequently spread to other centres and was used as a model in the formulation of the Native (Urban Areas) Act of 1923 which, in post-colonial society, was to be the principle instrument in administering, controlling and managing urban Africans.

Through its housing process the municipality gained control over its African population. Conditions in the developing but uncontrolled peripheral zone were different. The Durban housing process was directed at employed Africans and could not accommodate the inflow of a floating, surplus population, required, in part at least, as an accessible labour reserve. A permanent housing shortage and resort to informal solutions in uncontrolled space was the inevitable outcome of the contradiction. In this way the macro-structure of the colonial metropolitan area containing a controlled core city and an uncontrolled and problematic periphery was, in effect, predetermined. It is a structural characteristic of many Third World cities and is one which has never been overcome in Durban as later analysis will show.

The post-colonial city (1910–48)

For nearly 40 years after Union, South Africa's white group sought to entrench inherited relations of dominance in the social formation. Much was achieved at the national level but control over forces of metropolitan development proved only partially effective.

Expansion of economic activity and particularly of manufacturing industry from the mid-1930s gave rise to substantial growth in Durban's population as labour needs increased. Coincidentally, African rural-urban migration burgeoned as deteriorating economic conditions in the 'native reserves' caused a significant floating population to congregate in the metropolitan area. As the population grew it changed in composition and by 1951 Indians (170,000) and Africans (170,000) significantly outnumbered whites (149,000). A small marginalized coloured population of 17,000 people made up the total of 507,000 people.

Producer land uses developed a pattern characteristic of port cities in capitalist societies. Focused on the dual CBD, sectors of planned industrial land in white ownership extended mainly southwards on flat land associated with harbour facilities and transport routes. Holiday facilities and amenities were developed on the ocean front and, in outer-periphery suburbs, smaller concentrations of tertiary and industrial activities became foci of a multi-modal metropolitan structure. Extended over time, this basic structure remains characteristic of the contemporary city (Figures 5.1(b) and 5.1(c)).

Municipal government, an essential agent through which urban control is exercised, was first extended as late as 1932 when the inner periphery was incorporated into Durban as the 'added areas'. In the outer periphery, forms of municipal control emerged only from 1939. In both instances it followed, rather than accompanied, urban development and its tardy implementation was to contribute to the emergence of problems of control.

White planned residential areas, generated within a differentiated housing market, continued to expand without obstruction over the most desirable and accessible land. By 1948 they occupied the old borough, select sectors in the inner periphery and suburbs in the outer periphery.

Burgeoning African residential development presented a different picture. At the national level, comprehensive legislation intended to control African urbanization was progressively set in place. It was designed to promote low-income housing, prescribe the conditions and administration of African urban life, circumscribe African urban residential forms, ownership of land and mobility and to impose segregation. Municipal authorities served as agents of the state in administering the legislation. The problem lay in its permissive nature and, more importantly, in a lack of effective financial underpinning. Municipal responses were tardy and incomplete.

Holding the view that its African population was essentially migratory and floating, Durban continued to rely upon the extension of its established, economic, municipal, state and private single worker work zone barracks and domestic quarters for its African housing system. Living conditions were miserable but the system enabled effective control. By 1946 some 60 per cent of metropolitan Africans were resident in this way and concentrated particularly in the work zones of the Durban municipal area.

African female immigration had occurred from the 1920s, family formation increased and emphasis was increasingly being placed on the desirability of an African, family-based labour force for manufacturing industry. This issue was weakly addressed. By 1946 only two small, regulated rental housing townships at Lamont (1935) and Chesterville (1946) in the inner periphery had been established. They housed some 9,000 people or 7 per cent of the metropolitan African population.

Barracks and township housing continued to be mainly financed from Native Revenue Accounts that included profits from traditional beer sales. These sources proved inadequate in meeting real needs, though the system ensured that African housing was self-financing. Indeed by 1948 a commission of enquiry reported that 'the Durban Borough (i.e. the white ratepayers) had never been called upon to pay a penny towards African housing' (South Africa, 1948b, p. 93). The outcome was a rising incidence of informal African housing concentrated in the inner periphery and, particularly, at Cato Manor west of the old borough. By 1946, 33 per cent of Durban's Africans were living in deplorable circumstances, without effective water supplies or sanitation and insecurity of tenure became a source of tension and discontent.

In the outer periphery the freehold tenure African township of Clermont was privately established in 1931 precisely for that reason (Swanson, 1983). It too was largely informally structured, however, and by 1946 housed over 4,000 people. Urban freehold tenure, of course, violated the principle of temporary sojourner status and its practice for Africans was finally prohibited from 1937.

The majority of Indians were poor and more than half of them lived in slum conditions or shacks distributed mainly in the inner periphery and mainly segregated from whites. A smaller, more affluent proportion of the population, exercising its property market rights, continued to diffuse into inner-city white residential areas. Intense white opposition grew, grounded in perceptions of declining property values and cultural bias, and was the subject of an enquiry in 1921 (South Africa, 1921). Title deed clauses, preventing Indians from purchasing land, were introduced in 1922 and unsuccessful attempts were made fully to segregate Indians under the Class Areas Bill of 1924 and the Areas Reservation Bill of 1925.

By 1940 growth pressures arising from the African and Indian populations had mounted and it became evident that control over urban structure was failing. New measures to restore control were introduced. Two enquiries investigated ongoing Indian 'penetration' of white residential areas (South Africa, 1941 and 1943) and led first to the imposition of the Trading and Occupation of Land Restriction or 'Pegging' Act in 1943. It prohibited property transfers between whites and Indians for three years. The Asiatic Land Tenure (or 'Ghetto') Act followed in 1946 and, for the first time in Durban, official segregation was imposed upon Indians. 'Scheduled Areas', essentially confirming areas already in Indian occupation, were defined for ownership and occupation. The Act was not retrospective, however, and it did not eliminate the question of existing mixing or further mixing within scheduled areas.

Deteriorating control over housing emerged as a nationwide problem, and was addressed in 1944 by the establishment of the National Housing Commission. It provided for more effective access to housing capital loan funds to enable local authorities to fulfill their legal housing responsibilities for poorer groups under the Housing Act. Some 10 years were to pass, however, before the new funding mechanism was taken up in the development of low cost African housing in Durban. From 1945 Durban also became subject to the provisions of the comprehensively consolidated Natives (Urban Areas) Act.

Notwithstanding new and stricter control measures Durban remained a complex social mosaic in the late 1940s. In the white view its structure blurred; it contradicted fundamental relations of society and was redolent of incomplete control. The efforts of Africans and Indians to establish their urban status had produced forms of development which were cast by whites as 'problems'.

Segregation levels in 1951 were high: segregation indices between whites and Indians and Africans respectively, for example, were 0.91 and 0.81, where 1.0 represents absolute segregation (Davies, 1981). However, population distributions were complex, with growth not containable and in some cases obstructive to lines of white expansion (Figure 5.1(b)). Crowded and expanding inner-city Indian and coloured enclaves and mixing zones existed. High proportions of Indians lived in slum conditions. A rapidly expanding African population was accommodated and ineffectively segregated in working zones and in haphazardly scattered townships and unregulated shack concentrations in the inner-periphery, all perceived as serious health and social hazards. African influx control under permissive legislation had failed and control over peripheral space had faltered.

That circumstances such as these were also being in various ways experienced in other cities, became a national issue. That they were outcomes of structural forces created and maintained by the dominant group was not appreciated. Rather they became a major factor in the formulation of the more intensively coercive apartheid policy instituted under the Nationalist government from 1948. It was to be the ultimate framework of control to secure a formal city more effectively structured in the image of the social formation.

The apartheid city (1948–75)

The implementation of urban apartheid was a protracted and incremental process. An initial period of legislation and comprehensive, retrospective social engineering planning took up nearly 10 years. Group Areas were first proclaimed for Durban in 1958. The metropolitan area was to embody principles of sectoral design making effective use of physiographic divisions of its site in

creating segregated, consolidated and strongly defined Group Areas with growth
hinterlands for whites and Indians (Davies, 1981). Africans were to be accommodated mainly in the distant periphery to the north (KwaMashu) and south
(Umlazi) bordering on and within KwaZulu respectively.

The inner city core was characteristically reserved for whites and inner city
Indian enclaves, except for the Indian CBD, proclaimed as a 'controlled area'
for later decision, were eliminated. The zonal inner periphery was breached at
the expense of Indian and African occupied areas to provide extensive, unob-
structed sectors for white suburban expansion towards outer-periphery white
suburbs. Sectoral group areas for Indians were created over less desirable and
less accessible land already in their possession as 'scheduled areas' in remaining
parts of the inner and outer peripheries. The small coloured population was
allocated 'island' group areas in the inner and outer peripheries. (Figure 5.2(a)).

Asymmetrical relations in the social formation were deliberately reflected in
the core-periphery spatial relations of the design. In the spatial sequence of
group areas marginalized Indians and coloureds effectively buffered dominant
whites from subordinated Africans. Whites, on grounds of economic status,
were allocated a disproportionate share of space relative to population size,
and space that was desirable and convenient of access. Less than 2 per cent of
whites were to be affected by group areas disqualification and the economic,
social and psychological stresses arising from removals. Over 50 per cent of

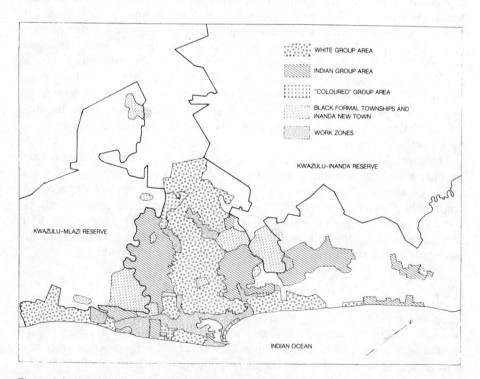

Figure 5.2a   Durban: metropolitan group areas outlines

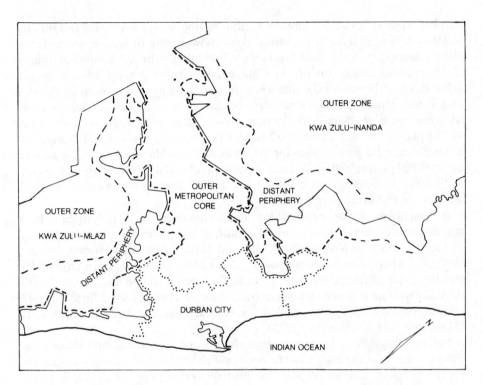

Figure 5.2b Metropolitan structural outlines in the Durban functional region

Indians and nearly 70 per cent of Africans were to be relocated in areas distant from major work centres in Durban situated mainly in the centre and south of the city. On economic grounds future development in white areas was to proceed uninterruptedly within the land market process. In Indian and coloured areas only approximately 40 per cent of future development could be expected to be market generated and heavy reliance rested on the state controlled, self-financing housing process to create rental and sale housing. Space allocated to Africans was to be state owned and the population taken up in an initially state funded but self-financing rental housing system. The form and qualities of mass formal housing for family and hostel accommodation would reflect an impermanent, environmentally deprived urban status and would be readily administered and policed. Access would be severely constrained by the need to possess legal rights to urban residence and African business formation would be severely curtailed.

The future structure of the metropolitan area was conceived entirely in formal terms and was intended also to facilitate and justify the imposition of separate education and health service systems. Levels of access were to vary sharply.

Over a period of 10 years the metropolitan area, and in particular the Durban municipal area, experienced uncertainty and crisis management as formal municipal initiatives were inhibited. Serious rioting between Africans and Indians in 1949, arising at least in part from African insecurity of tenure, exacerbated circumstances. By 1950, half Durban's Africans were living in inner-periphery shack concentrations, mainly in Cato Manor, housing some 68,000 people (University of Natal, 1952). Shacks were also evident in the south of the city and in municipally uncontrolled edges of the outer periphery at Clermont and in areas south of Pinetown. Durban built a large new single worker hostel (4,272 places) in 1951 and in 1952 (under the Squatting Act, 1951) established a controlled emergency camp for Africans at Cato Manor. By 1957 it housed over 90,000 people and 30,000 more clung precariously to its fringes in uncontrolled shack housing. At least 10,000 Indian households were also shack dwellers in the inner periphery.

Restructuring of the metropolitan area commenced in 1958. African shack concentrations were progressively demolished and their populations rehoused in the townships at KwaMashu (1958) and Umlazi (1962) as housing became available. Klaarwater township was established south of Pinetown from 1960. By 1965, the African removal process had been completed. Some 95,000 Africans with legal urban status had been relocated but at least a further 30,000 people, probably 'illegally' resident in Durban, had not been accounted for (Maasdorp and Humphreys, 1975).

Indians disqualified by group areas design and who could afford to do so were forced, often at high cost, to develop private housing in areas designated for the purpose. For poorer Indians, probably 60 per cent of the total, public housing, mainly for home ownership, was generated from the late 1950s first at Merebank (south) and from 1963 at Chatsworth. In keeping with the structural relations of the group, housing services and environmental qualities were superior to those adopted for African township development. By the mid-1970s, 25,000 dwellings had been built, 20,000 in Chatsworth alone (Corbett, 1980). Since 1975 development has been concentrated in northern Indian group areas at Phoenix and Newlands where a further 20,000 dwellings have been completed to house over 100,000 people.

Though much remained to be done, by 1964 it appeared that formal control over metropolitan space would be achieved. Durban was no longer to accommodate informal development and its subordinate African population was under effective administrative management and constraint. By 1970 the census showed that the metropolitan population stood at 1,053,981 people made up of 267,760 whites (25 per cent), 407,896 Africans (39 per cent), 333,089 Indians (32 per cent) and 45,236 coloureds (4 per cent). Segregation indices between population groups stood at near absolute levels (Davies, 1981) and segregated social systems functional to urban apartheid were being set in place. It appeared that provisions for future growth could be accommodated by incremental development. While that image proved to be valid for whites and to a substantial degree also for Indians and coloureds, as their economic mobility in the labour force improved, it was an illusion in the case of Africans.

Low-income Indian housing mainly directed at those disqualified by group areas design, failed to meet the full formal housing need. Shortages, which today stand at 31,000 dwellings, remained. Overcrowding and backyard shack housing have become characteristic.

For Africans, planning data grossly underestimated real numbers (by up to 50 per cent in some cases) and, as they attempted to urbanize, evasions of influx control regulations were high. Such people tended to filter into formal townships raising levels of overcrowding or continued to settle informally in areas lacking effective municipal control in the edges of the outer periphery. In 1965 at least 5,000 shacks continued to exist, housing some 38,000 people. That they were scattered in remote places far from view was comforting for whites but indicative of an intractable on-going situation of incomplete control.

The situation worsened in response to several interrelated factors. From 1960 the state introduced its policy for territorial apartheid (separate development) and the creation of 'bantustans'. Political, economic and social containment of urban Africans, influx control and limits on African metropolitan labour growth, were to be balanced (in theory) by decentralized economic development and settlement about and within bantustans – in the Durban case, KwaZulu. In consequence, state capital funds for metropolitan African housing were severely limited from the late 1960s. In the Durban case only some 5,000 formal African family dwellings were built in the period 1970–86 in new townships at Ntuzuma (north), KwaNdengezi (west) and KwaMakutha (south). A major hostel, housing 10,000 single workers was built at KwaDabeka (Clermont), however, and hostels at Umlazi were extended.

Decentralized development was costly and largely failed to achieve its objectives (Bell, 1973; Davies, 1986). In the face of increasing African rural worklessness and poverty, urbanization was the logical solution. Urban influx control was resisted. In Durban the proximity of the KwaZulu boundary and the presence of released land, defined in terms of the 1936 Native Trust and Land Act, and already largely in African occupation, was to be crucial in that process.

Relatively effective control over African influx could be maintained in formally administered townships and hostels but not within KwaZulu, or on released land, and only with difficulty in other less controlled peripheral localities. On the edges of KwaZulu and on released land it was possible for urbanizing Africans to attain legal tenure to land in a variety of ways and to gain access at least to opportunities for urban employment. Legal constraints on entry to formal employment in Durban made the process difficult, however, and a resort to informal activity was a necessity. Rapid informal growth took place in what had become a new distant metropolitan periphery beyond the control of established urban local authorities (Figure 5.1(d)). The peripheral informally settled population rose from 38,000 in 1965 to 150,000 in 1969 and 275,000 in 1973 (Maasdorp and Humphreys, 1975).

As settlement spread outwards the structure of metropolitan space progressively straddled the boundary of KwaZulu northwards and southwards. From as early as 1973 (Natal Town and Regional Planning Commission, 1973) a broader Durban Region, now defined as the Durban Functional Region (DFR), became recognized by planners as a logical planning unit. It consists of the inner, formal metropolitan core, a largely uncontrolled, informally urbanized distant periphery and an outer zone of future urbanization containing a population functionally linked by employment and services to the metropolitan economy. A now familiar core-periphery metropolitan structure had resurged (Figure 5.2(b)). As in the past it distinguishes between areas of formal and informal development and groups of haves and have-nots.

The total population of the DFR (including the outer zone) was estimated to be 2,567,000 people in 1985 made up of 1,575,000 Africans (61 per cent), 576,000 Indians (22 per cent), 352,000 whites (14 per cent) and 64,000 coloureds (2 per cent) (Sutcliffe and McCarthy, 1989). By 1985 the metropolitan core accommodated 62 per cent of the total population and the distant periphery and outer zone 38 per cent. By that year, however, the distant periphery and outer zone contained approximately 972,000 Africans or some 62 per cent of the total African population of the DFR settled mainly under informal circumstances (Sutcliffe and McCarthy, 1989). By 1990, the Urban Foundation (1990c) estimated an informally settled population of 1.7 million in greater Durban.

Metropolitan core space is structured on 44 municipal authorities ranging in status from Boroughs to Development Service Boards, five African Town Councils in Natal and six African Town Councils in KwaZulu. It contains the work zones of the region, its major amenities and services and the formally structured, administered and serviced white, Indian, coloured and African residential areas and townships spatially segregated by group areas design.

Though averaged socio-economic data blur variations within groups, they provide an indication of the relationships in the social formation in the core area. In 1985 average annual household incomes of whites, Indians and Africans ranged from R31,000 through R14,000 to R8,000. Whites, forming 14 per cent of the population, earned 54 per cent of the income in the DFR. By contrast the ratio for Indians was 22 to 22.5 and for Africans 61 to 23.5 per cent. Of African households in formal townships, 13 per cent fell below a household subsistence level and 34 per cent of adults were not working (McGrath, 1989).

In sharp contrast the distant periphery accommodated no less than 40 informal settlements without formal local authority structure. The average annual household income in the distant periphery and outer zone in 1985 was only R4,860 and though these zones contained 38 per cent of the total population their people earned only 5 per cent of the total household income of the DFR. Approximately 63 per cent of adults were not working and 40 per cent of the population in the distant periphery had incomes that fell below the Household Subsistence Level (McGrath, 1989). The importance of informal activity as a means of survival is clearly crucial. These areas furthermore have a weak physical and social infrastructure and there is a tendency for dense shack concentrations to cluster near formal townships where services are available and transport accessible. For those in employment, journeys to work, often indirect, consume between two and three hours and are costly in money and effort.

In these developments lay evidence of the failure of urban apartheid to structure and maintain a formal city form and to control irresistible forces of urbanization which, in the South African context, represent forces of resistance to imposed containment. That such forces became coupled to a complex of other forms of resistance nationally gave rise to the emergence of a conflict threshold evident from the mid-1970s (Davies, 1986).

The city in transition: 1976 and thereafter

The State's response to structural conflict was to introduce multifaceted, adaptive reform but not to penetrate fundamental relations of the social formation. On the one hand, it was designed to reduce constraints experienced by marginal and subordinate groups and to raise levels of incorporation to diffuse forces of conflict. On the other, it has emerged as a means through which the State might manage, rather than coerce, processes of urbanization and urban development, and involve private initiative and financial resources in housing delivery and the provision of services. In important respects it has been beneficial to dominant white economic interests and in some cases it has been manipulated to retain ultimate white control. Fundamental geographical imperatives expressed for example in group areas and in the separation of systems functional to apartheid like education have remained intact.

Centrally determined forms of change in Durban are in many ways shared by other cities. They do, however, find individual expression in the particular local context. Change affecting the marginal Indian group has included the proclamation of the Indian CBD as a group area in 1975 and, in 1986, of the entire CBD as a 'free trade' area. In an environment of change, group area violations have frequently been overlooked. Residential filtering into some white residential areas by Indians, coloureds and some Africans, induced by housing and land shortages, has taken place. Inner city rental flats and boarding houses and some suburbs bordering Indian Group Areas such as Westville have been most affected. The extent of the process remains unmeasured at this stage. To divert the process, however, two small free settlement areas have been proclaimed in Durban since 1988 in the Warwick Avenue triangle and at Cato crest in Cato Manor.

These changes have been economically and politically beneficial to Indians and coloureds. White businesses too have benefited as has the city itself, through its tax base, by the ongoing vibrancy of the CBD, appropriate occupation of space and political acquiescence. Conflict over the urban environment involving the marginal groups has been significantly shifted away from whites.

Bearing its specific development history in mind, change affecting African urban settlement was largely pre-empted in the Durban context. That the majority of Africans lived in formal townships and informal settlements in KwaZulu or on released land meant that they enjoyed access to tenure prior to the grant of ownership rights to urban Africans in 'White South Africa' in 1978. The majority of Durban Africans had also evaded influx control measures by settling in the edges of KwaZulu and townships it controlled long before that constraint was abolished in South Africa in 1986. Those areas, however,

had absorbed informal housing development and in essence released the metropolitan core and the State of their housing and services responsibilities. That
factor has probably also underlain a very tardy response by the state in
providing new land and urban infrastructure for African housing on the Natal
side of the KwaZulu border for many years.

The State, through its Department of Development Aid, however, was
nevertheless constructively involved in an important, experimental site and
service housing scheme established in co-operation with the Urban Foundation
on released land in the northern distant periphery from 1983. The scheme, at
Inanda New Town, integrated planning and infrastructural resources of the
Department, advice and financial resources accessible to the Foundation, community participation and individual involvement in privately creating informal
and formal housing. The scheme now houses some 40,000 people. It was a
forerunner of the State's concept of 'orderly urbanization' and an important
indicator of a future urban development form. It also demonstrates management
of African urban location and the co-optation of a local community in an area
of severe conflict (Sutcliffe, Todes and Walker, 1989).

Heightened African mobility has, within the severe constraints of a locally
over-supplied unskilled labour market, at least granted an opportunity for
work seekers to search for formal employment. Furthermore, the accommodation of controlled informal activities in the metropolitan core, on the ocean
beach front for example, and unrestricted participation in African townships
and informal settlements, now provides a means of survival, notwithstanding
well known structural exploitation which informal entrepreneurs may experience.

Attempts to impose a new local government structure in Durban from 1982
based on group autonomies and using existing Indian and coloured group areas
and African formal townships as a spatial framework, have been largely unsuccessful. Indians and coloureds have rejected the proposed system on grounds
of perpetuated discrimination and white manipulation of urban space to entrench
urban apartheid. African Town Councils have been established but the legitimacy
of the imposed system has been called into question and is a major source of
community conflict. The administrative divide between Natal and KwaZulu
furthermore has enabled KwaZulu to reject the imposition of a proposed
Regional Services Council for the DFR. That white local authorities in the
proposed region represented on the Council would hold decisive power on
grounds of their volumes of services used, is a major factor in that decision. An
alternate Joint Services Board somewhat more democratically structured has
been negotiated between KwaZulu and Natal.

Relaxation of constraints on access to amenities has in general produced only
a minor ripple in Durban. The opening of its nationally important bathing
beaches, however, has been problematic in its initial stage. This is a sensitive
sphere and overcrowding and behavioural problems in the high summer season
of 1989 gave rise to a significant withdrawal of white holiday custom. Should
the situation continue, serious economic and employment consequences could
arise. Durban has made major improvements to its beach infrastructure to
cope with the problem but it remains an indicator of the broader national issue

of the need to attain attitudinal changes in inter-ethnic relations in an environment of transition.

In the operational sphere, new freedoms of association have generated high levels of interaction in which economic and social issues and on-going containment of political power relations have become major foci of organization by a bewildering array of interest groups and contenders. Tensions and uncertainties in the city have risen and are evident in frequent street gatherings, rallies, demonstrations and marches, now officially permitted. Fundamentally directed at attaining structural change in the relations of the social formation and distribution of political power and resources, strong differences of principle and approach have emerged between contending groups. In that context the DFR has become an extraordinary arena of conflict, now frequently expressed in physical violence.

The conflict is presently centred, though not exclusively so, in a power struggle between African groupings opposed to the prevailing social order but differing on matters of principle and means of attaining change. At the core are progressive groups (UDF, ANC and COSATU) and the traditionally rooted Inkatha organization. Beyond that major struggle are less directed forces of conflict. They arise from local struggles over scarce resources and particularly the control of land, 'war-lordism', mobilization of frustrations of a deprived, often unemployed but politicized youth, unemployment, poor housing and poverty, tensions of urbanization, exploitation of opportunities to settle factional or personal scores and crime. Collectively they constitute a major urban crisis which, in as much as it is characteristic also of the Pietermaritzburg region (Chapter 6), has meant that a state of emergency proclaimed in 1987 continued to be operative in Natal. In August 1990, following talks with the ANC, the government agreed to consider lifting it and did so in October.

Thus far concentrated in the incompletely controlled distant periphery and African formal townships, the conflict plays itself out beyond the experience and out of sight of the white metropolitan core but has marginally affected neighbouring Indian groups. Its disruptive and frequently tragic effects on individual lives is traumatic, economic effects will be material and psychological impacts upon the city in general have become tangible.

The future

The complex structure of contemporary Durban, now suffused with crisis, is the climactic outcome of a system of social relations which has, in a convoluted history, contained the integration and incorporation of the majority of the people. Inherent forces of conflict, now evident nationally, have finally shifted society into a phase of structural transition.

The central concern in South African society in 1990 has become the creation of circumstances favourable to negotiations between the existing state and opposing forces of resistance over future political relations. By extension future relations of the social formation as a whole will be in contention.

The current position is uncertain but successful negotiation would lead at least to the abolition of ethnicity as a primary reference in society. Pragmatism suggests also that a new society is likely to remain based in a mixed-capitalist mode of production, possibly adapted in ways to raise the competitive opportunity levels of Africans. Important implications for future urban structural relations arise.

A vital initiative would be a search for an administrative structure which could appropriately integrate the space of the DFR to overcome current discriminatory fragmentation and exclusion. Compositions of existing local authorities would change. Whether integration would necessarily result in significant shifts in resource allocations to address existing, stark disparities in qualities between areas, however, is another matter. In a new urban environment where economic rather than ethnic differentiation will be a primary reference, economically mobile Africans in particular may not necessarily favour wholesale redistribution of local authority resources as the experience of Harare suggests (Chapter 13). Sharp socio-economic, amenity and service disparities are thus likely to persist despite administrative integration.

In a city like Durban the release of ethnic constraint on public sector employment and a shift in the employment status of a significant number of formally qualified Africans could be an important development. Economic mobility in the private sector is likely to be more constrained. Entry to higher status employment is likely to be more strictly controlled by productivity potentials of entrants including their education, training and experience and by established networks of association. Ownership of the means of production, however, is likely to remain vested mainly in white hands in the short to medium term at least.

Improved economic mobility of a significant number, but small proportion, of Africans will encourage residential diffusion into former and more conveniently located exclusively white residential areas in particular. Central city rental flats and middle- and lower-middle income suburban areas are likely to be most strongly affected but diffusion will nevertheless be widespread and not clustered. Integration of functional systems such as school education would be an automatic outcome of the diffusion process. Shortages of land in all areas presently occupied by Africans will be a major push factor in the diffusion process. Were large-scale white emigration to occur, property values are likely to remain static or marginally decline as blacks take up vacancies. The reverse would be true if white emigration were to be low.

Further relaxation of present urban structures is likely to be severely constrained, however, by basic development realities. Durban's population, increasing at a rate of 4.5 per cent per year will grow from a presently estimated 3.4 million people to an estimated 5 to 6 million people by 2,000 (Sutcliffe and McCarthy, 1989). The proportion of whites will have fallen to 8.3 per cent, Indians to 13.7 per cent and coloureds to 1.5 per cent but Africans will comprise 76.5 per cent of the total. The African population will remain youthful with progressive growth tendencies.

Given a modest outlook, the economic growth rate will be below that of population increase and high levels of unemployment and poverty (mainly

among Africans) will be an outcome, notwithstanding possible high.. .. ,
ment rates in the public sector. If formal housing for all is taken as a desirable
goal, present housing shortages estimated to be 31,000 for Indians, 4,000 for
coloureds and at least 160,000 for Africans will be exacerbated (Sutcliffe,
1989). The outcome for the majority of Africans will be a resort to informal
housing and a perpetuation of the circular relationships of poverty and weak
economic mobility.

The spatial structure of the DFR including its highly structured residential
relationships, which reflect sharp economic divisions, is likely to remain intact.
Spatially distorted workplace and residence relations and administrative and
service fragmentation will remain sources of development problems. Informal
settlement in the distant periphery, particularly in areas accessible to neighbouring service infrastructure, is likely to intensify and, in the light of existing
patterns of violent conflict, be an ongoing source of crisis. An overspill of
informal illegal squatter settlements into large vacant areas and fringes of the
formal metropolitan core cannot be ruled out.

Development needs stand out starkly. Any attempt to meet these needs
requires, *inter alia*, political stability, education to raise levels of development
potentials, appropriate housing delivery systems, investment in job creation
and in particular in labour intensive small and informal businesses, service
infrastructure, creative co-ordination of state and private sector development
resources and encouragement of individual initiative.

Future Durban will clearly be an African Third World city with all the
structural difficulties that status implies. It does not mean, however, that the
prospect is necessarily bleak. Very positive forces in South African society
could be harnessed. The local region, Natal and the country as a whole, in the
African context, are extraordinarily well endowed with resources and already
possesses a well developed, if distorted, economic, social and physical infrastructure in which Durban shares. Within this context, development directed at
creating means through which the potentials of the hitherto ineffectively incorporated African population could be released, and initiative raised, could
materially alter the parameters of its urbanism. This is the essential task which
awaits the new society.

6
PIETERMARITZBURG
T. M. Wills

Fox Hill stands to the south of Pietermaritzburg, and from its crest the full extent of the metropolitan area is revealed in one panoramic sweep. To the north, south and east lie the city's suburbs. In all these areas the boundary of the city is well defined; one can detect a hard edge where suburbs (or industrial zones) end and farmland or forest begins. This is not true of the western sector of metropolitan Pietermaritzburg however, where the African residential areas stretch up the Umsindusi valley, with the rigidly planned formal townships giving way to more amorphous and informal suburbs (Figure 6.1) as one moves away from the city centre, and enters an outlier of the fragmented bantustan of KwaZulu. The African commuting belt stretches, in fact, some 30 kilometres, at which point the density of settlement has reduced considerably and a more traditional landscape pattern is evident. Over half of the metropolitan population of approaching 500,000 is housed within these townships and 'peri-urban' areas. It is within these areas that political violence has become endemic, with clashes between Inkatha and the 'Mass Democratic Movement' having cost over 500 lives since 1987 (Niddrie, 1988; Cross *et al.*, 1988).

From the Fox Hill vantage point one can also clearly see that the suburbs once designated African, Indian or coloured are detached from the core of the city and the flanking white suburbs by strips of open space or industrial areas. This apartheid city form has been maintained for decades by a plethora of laws governing where people may live and move, but has its origins in the founding of the town by the Voortrekkers and its subsequent growth as a colonial capital under British rule. The 'colonial' character of Pietermaritzburg did not disappear with Union but was enhanced during a period of social engineering which preceded the coming to power of the National Party in 1948, and the subsequent imposition of a national urban model based on racial separation.

Less clear from this vantage point are the signs of change: the beginnings of Pietermaritzburg's transformation to a post-apartheid city. Although fundamental spatial and social change will only be evident when the Group Areas Act

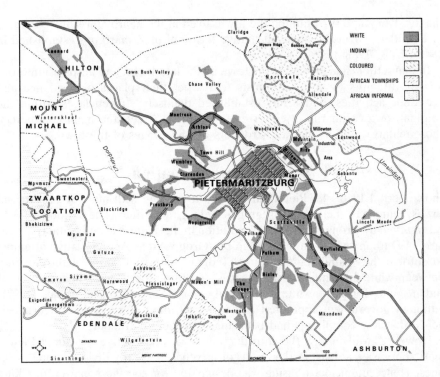

Figure 6.1   The racial geography of metropolitan Pietermaritzburg, 1990

and other remaining discriminatory legislation is repealed and new local government structures put in place, the city is inexorably changing as the sharp edges
of the apartheid city are blurred by the sheer momentum of urbanization.

The colonial capital

Pietermaritzburg owes its existence to the Voortrekkers, who, in escaping
British domination, were determined to establish an independent republic in
what was to become the colony of Natal. The Voortrekker 'dorp' of Pietermaritzburg (Haswell, 1979) was to be the capital, and as a consequence was
laid out in 1838 on what was, in those days, a grand scale. The dorp was
intended to be the home of Voortrekker families and their servants, and no
provision was made for the accommodation of the local indigenous population.

  The short-lived Republic of Natalia was soon replaced by the colony of
Natal, and Pietermaritzburg became the colonial capital, acquiring by Union in
1910 all the attributes of a British colonial town in Africa. A garrison was
stationed at Fort Napier, overlooking the original Voortrekker grid within
which the white colonists lived. African residents (other than domestic servants)
had been effectively barred from the town through by-laws prohibiting the
construction of wattle and daub buildings and other traditional building styles.
If not occupying rooms in the 'barracks' set up for workers, they squatted on

the low-lying land surrounding the town, lived in the nearby mission settlement of Edendale, or in the adjacent 'native location' of Zwaartkops established in 1846 (Wills, 1988a).

Indian settlers had arrived in the town by the late 1860s, and despite numerous *ad hoc* attempts by white colonists to restrict their places of residence and economic activities, had firmly established themselves in enclaves within the colonial town (primarily at either end of Church Street, the main commercial thoroughfare), and as market gardeners on the fringes of the town.

Post-union segregation

In the period 1910–48, Pietermaritzburg, in keeping with most South African towns, came to be increasingly segregated along racial lines. The period 1910–46 was marked by strong 'anti-Asiatic' agitation in the Transvaal and Natal (particularly in Durban) which culminated in pre-Group Areas Act legislative attempts to enforce the spatial segregation of Indians. The root cause of the hostility between white and Indian residents was said to be the alleged 'penetration' of Indians into formerly white residential areas (Maasdorp and Pillay, 1977), although government-appointed Indian Penetration Commissions heard evidence that few complaints had been made in Pietermaritzburg.

However, well before the introduction of any restrictive measures Pietermaritzburg was, at least in the minds of its officials, clearly divided along racial lines. Hence municipal housing was designated for specific racial groups long before the Population Registration Act (1950) made such racial division mandatory, or the Group Areas Act enforced territorial separation.

It has been generally accepted that the first legislation aimed at residentially segregating Indians in Natal was the 1922 Durban Land Alienation Ordinance No. 14, permitting the Durban City Council to include an 'anti-Asiatic' clause in the title deeds and leases of Borough land (Maasdorp and Pillay, 1977). However, there is evidence that Pietermaritzburg had taken the initiative some decades earlier, and by 1898 the City Council was inserting such clauses into the conditions of sale of townlands (Wills, 1988a).

In response to the passing of the Natives (Urban Areas) Act in 1923, a landmark in the segregation of Africans within urban areas, the Mayor of Pietermaritzburg made the following comment:

> The question of native housing in the borough has for many years been one of the thorniest description. A native location or village has been discussed for years, but it has hitherto been impossible to do anything because of the objections of burgesses to allow the village to be formed in some particular part of the borough or the other. The passing of the 'Urban Areas Act' makes it incumbent upon local authorities to deal with the question or the Minister of Native Affairs can step in and apply compulsion.
> (*Pietermaritzburg Corporation Yearbook*, 1924, p. 33)

Within seven years the entire borough was subjected to the provisions of the 1923 Act, which also directly influenced the City Council to undertake its first major public housing scheme for Africans within the borough: namely Sobantu

Model Native Village (Figure 6.1). Prior to this, local authority involvement had ended at providing barracks for, amongst others, *togt* (casual) workers, Africans in municipal employment and rickshaw pullers. A major impetus for building such a village came from the Medical Officer of Health who firmly believed that the slum conditions that had developed in peripheral shanty settlements could not otherwise be eradicated. The 'sanitary syndrome' (Swanson, 1977; Beavon, 1982) reared its head again when the site for the village was decided upon by a plebiscite of white voters, who selected a site not favoured by African residents because of its proximity to the municipal refuse dump (Peel, 1988). A sewerage treatment plant was a later neighbour.

Thus by 1950, when the Group Areas Act was first promulgated nationally, Pietermaritzburg was a racially divided community. White suburbs had spread out north-west and south-east from the central area (the original Voortrekker grid), while the bulk of the African population lived outside the borough, south of the city in the valley of the Umsindusi River. The exceptions were Sobantu village and a few landowners on semi-agricultural land dotted around on the periphery. The Indian and coloured populations of Pietermaritzburg were concentrated in the central area, with smaller pockets of Indian and coloured landowners also dotted around the periphery, particularly in areas where market gardening was possible. Within the central area, tight-knit Indian communities had evolved a distinctive townscape with workplace, mosques or temples, schools and homes in close proximity. Effectively two central business districts (CBD) had evolved; a 'white' CBD, patronized by all, but developed in the mould of its European masters, and an adjacent 'Indian' CBD whose townscape and patronage were distinctly different (Haswell, 1985).

It was not until 1960 that group areas were defined in Pietermaritzburg. The 1950s were marked by uncertainty and anticipation. Witnessing the changes being wrought in other cities within which group areas had been demarcated at an earlier date, plans were made by the municipality with the encouragement of the State to construct new public housing schemes for Indians, Africans and coloureds on the fringes of the built-up area. Despite having grown to be an integral part of the city, the future development of Sobantu village was curtailed by central government edict, and the inhabitants remained under threat of removal until reprieved in 1979. With the passing of the Group Areas Act it became clear that it was racial *separation* that was sought, a racial rationalization, and not just segregation, which had in any event characterized South African cities since Union (Christopher, 1987a).

The decade following the implementation of group areas proclamations for Pietermaritzburg in 1960 coincided with a brief period of rapid industrial growth, following the declaration of the city in 1963 as a Border Area in terms of the government's industrialization and bantustan development policy. Generous concessions were offered to industrialists willing to relocate, land sales boomed and population growth accelerated. However, in 1970 the Border Area concessions were removed, and industrial growth tailed off. The black housing backlog generated during this brief period is an enduring legacy, because unlike industrial concessions which could be granted or withdrawn at will, the process of urbanization initiated by this artificial stimulation of the local economy could not as easily be halted. Much of the population growth took place outside the borough of Pietermaritzburg, in areas administered by KwaZulu or directly from Pretoria (Wills, 1988a).

The proclamation of group areas in Pietermaritzburg

It is not known with any certainty exactly how many people were forced to relocate as a result of the implementation of the Group Areas Act in the city (Figure 6.2), but various estimates were made at the time. Motala (1961) suggested that about 900 properties were affected, involving over 9,000 people, while municipal estimates were that 700 properties and 15,000 people would be affected (*Pietermaritzburg Corporation Yearbook*, 1960/61). The Indian community bore the brunt of the changes demanded by the Act, accounting for 76 per cent of those moved. Eleven per cent of the properties affected were owned by coloureds, while a further 12 per cent were African-owned. Only one per cent of the properties were white-owned.

Unlike their Cape Town counterparts, the Pietermaritzburg City Council co-operated to the extent that it drew up proposals for the racial zoning of the city (in conjunction with the Natal Local Health Commission). These proposals, and those put forward by a State-appointed Reference and Planning Committee, together with the counter-proposals proffered by the Natal Indian Organization, were considered by the Group Areas Board (Horrell, 1956). The Reference and Planning Committee's proposals were the most radical, but formed the basis of the group area zonings finally accepted by the Group Areas Board.

The impact of the Group Areas Act

The allocation of space, although in many respects reflecting the historical pattern of settlement, was anything but equitable (Table 6.1). Recent figures show that the white community, forming 36 per cent of the borough population, have enjoyed access to 79 per cent of the land within the borough, while the Indian population (around 43 per cent of the present population) were allocated a scant 13 per cent of the land.

Within the Indian community particularly, the dire shortage of land available for private development has added an artificial premium to the market value of available land and property, exacerbating the dichotomy between public housing and the private sector, and leaving a growing number of middle-income families trapped between State-subsidized housing which they may no longer qualify (or wish) to occupy and private housing which they cannot afford.

The most select Indian residential area, Mountain Rise (Figure 6.1) lies within a horse-shoe shaped zone of industrial land near the city centre, and is subjected to high levels of noise and atmospheric pollution. Paradoxically land values on average are higher in this ecologically unattractive Indian suburb than in any of the very attractive peripheral white suburbs.

Perhaps the two most obvious changes wrought to the townscape of Pieter-

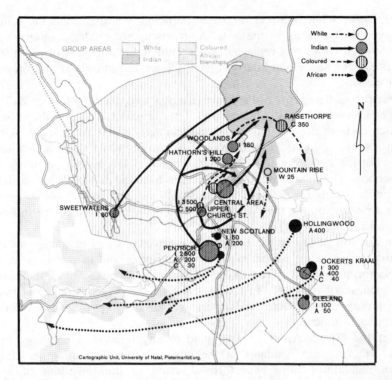

Figure 6.2   The group areas in Pietermaritzburg and the major relocations of population

Table 6.1   Land allocation, population and housing supply in Pietermaritzburg

Population group	Group areas (hectares)	%	Population 1988	%	Housing supply (Current)
White	9310	79	65080	36	Oversupply
Coloured	818	7	16730	10	Undersupply
Indian	1612	13	78260	43	Undersupply
African[1]	85	1	19180	11	Acute undersupply

Notes
[1] Borough figures alone. African population of metropolitan area is estimated to be an additional 320,000.
(*Source*: Municipal Estimates, 1988.)

maritzburg were the clearance of the Upper Church Street area of Indian and coloured residents and traders, and the elimination of the suburb of Pentrich (Figure 6.2). In the former case hundreds of residents were forced to give up their homes, many having lived above or behind their business premises. Nearly 50 traders were disqualified from serving a market they had come to dominate. More important than numbers, the townscape that had been created in the Upper Church Street area was particularly Indian, and the removals destroyed the character of the area. A streetscape of modest, colourful, small family-owned shops was replaced by an anonymous amalgam of chain stores, with no longer needed and derelict flats, school and mosque a mute reminder of the past.

Since the earliest years of Pietermaritzburg's existence the Camps Drift area had been home to many Africans and, later, Indians. By 1960 the suburb of Pentrich had developed, and was home to nearly 2,500 Indians, 200 Africans and a few coloured and white families. Not only did all the families have to move, but the enforcement of the Act eliminated seven shops, a mosque and four schools. The suburb literally disappeared from the map of Pietermaritzburg, and, until recently redeveloped as industrial land, survived as part of an open buffer strip, separating the African residential suburbs from the main body of the town (Wills, 1988a).

The division of the city into separate racial zones was unambiguous in the suburbs, with natural boundaries, transport routes and in one instance cemeteries, forming the required buffer strips between races, and enhancing the 'separateness' of daily life. In the central area the divisions have not been as clear, however, and white, Indian and coloured properties shared common boundaries in places.

After the Group Areas Act put paid to any hope of expanding Sobantu Village, the African residents of the city who qualified in terms of influx control regulations were henceforth diverted to the townships of Imbali, Ashdown or Edendale, all on the southern periphery of the city. Those who did not qualify were forced into the twilight world of lodgings, or into the peri-urban areas under African tribal control. Dependence on public transport thus became a way of life (Chapman and Cristallides, 1983; Voges, 1984). The heavily subsidized bus transport system needed to prop up the 'apartheid city' has been the target of frequent consumer boycotts and strike action, while the political violence that has beset the African residential areas of the city during the period 1988–91 has frequently culminated in the withdrawal of bus services, leaving the poor effectively stranded.

A more permanent threat to bus transportation comes from the rapid proliferation of mini-bus, or 'Kombi' taxis (Wills, 1988b). Spawned by discontent with the inefficiency of existing bus services and a spate of bus boycotts, these taxis carry about 60 per cent of the daily commuters into and out of the city. The majority are illegally operated, and it is estimated that between 1984 and 1990 the number of so-called 'pirate taxi' operators in Pietermaritzburg rose from around 200 to over 600. The taxis serve not only the African residential areas, where the poor quality and steep gradients of roads and a relatively low density of settlement suit mini-buses rather than buses, but the Indian and coloured areas as well. The popularity of the 'Kombi' taxi has geographical significance as well, because it has freed black commuters from the spatial constraints of a restricted public bus or train transport network. As a consequence larger areas of the city are now accessible, for example for shopping.

The chronic shortage of formal housing opportunities in the city for Africans is reflected in the proliferation of backyard 'shacks' and informal extensions to houses in townships such as Sobantu and Imbali. Municipal estimates in 1985 put the number of formal houses in Sobantu, for example, at 1,098 and the number of backyard shacks at 981. These *ilawu* or *mjondolos* as they are known, house family members or lodgers taken in to supplement incomes. The advantage to the occupants of these 'infill' shacks is that, unlike the peripheral informal settlements, a range of urban amenities is available to the lodgers as well as to the permanent residents.

The central area of Pietermaritzburg (i.e. the original grid layout) is home to about 23,000 people. Nearly half live in flats, but the central area does contain tracts of single-family housing in sound to excellent condition. Overcrowding reaches a peak, and physical condition a low point, in the comparatively small areas zoned for Indian and coloured occupation. Here homes compete for space with shops, warehouses, light industry, schools, temples and mosques. For many white households the central area provides a temporary home or cheaper accommodation than the suburbs. Others, however, are attracted by the historic character of the core of the city and its many fine old red-brick colonial homes on tree-lined streets.

Pietermaritzburg in transition: the open city debate

The issue of opening up all or part of the city to all races was hotly debated in local government circles in Pietermaritzburg in the five-year period leading up to the announcement in 1990 that the Group Areas Act was to be repealed. Prior to the announcement of the Free Settlement Areas Bill in 1988 the suggestion was made by a group of white city councillors that small portions of the city be designated as 'grey' areas, or areas of 'local option' where people of any colour would be able to reside and trade. The City Council in 1986 rejected the proposal, acting on the advice of the City Engineer and his planners, who recommended opening up the entire city, arguing cogently that piecemeal reform would be counter-productive, and might result in the very invasion and succession scenario, and consequent downward socio-economic spiral, feared by white (and many Indian) residents (Schlemmer, 1986).

In the 1988 local elections, however, the balance of power in the council swung towards the National Party, and when the City Engineer recommended that the entire city be put forward for designation as a Free Settlement Area in 1989, the proposal was rejected. What had been ignored by the liberal proponents of 'opening' the city was that in Free Settlement Areas a common voters roll must be established, and only advisory committees elected. White voters electing not to support the Free Settlement initiative (presumably largely right-wingers politically) could retain their right to elect representative bodies which would effectively then govern the 'free' settlement area. In 1989, however, the General Election saw a swing back to the 'left' in Pietermaritzburg. The mayor of the city then took the unprecedented (in South Africa) step of announcing a referendum of local voters (white, Indian and coloured) on the question of 'opening the city' to all, including Africans who would have been excluded from the vote.

During 1987 and 1988, while local politicians debated desegregating the city, the attitudes of Indian and white households were investigated in two preliminary questionnaire surveys of stratified random samples of 310 and 290 households respectively. The results suggested (Table 6.2) that although in both cases there was some resistance to racial integration, white households were much less prepared to give up their exclusive hold over sections of the city. In the 1988 study, in fact, 47 per cent of the white households interviewed said that they would seriously consider moving out if their neighbourhood were opened to all.

Perceptions of the impact that repealing the Group Areas Act would have on the property market differed significantly. For example, 48 per cent of the white households expressed the view that property values would drop in white areas after the repeal, whilst 50 per cent of Indian households believed that property values in white areas would increase. There seems to be a general white fear of negative impacts on property values (Christie, 1987).

Grey areas and change

In keeping with trends in other South African cities, so-called 'grey areas' began to emerge in Pietermaritzburg prior to the announcement that the Group Areas Act was to be repealed and despite the establishment of a 'Directorate of Area Management' in 1989 to investigate complaints of contraventions of the Act. Indian and coloured enclaves within the central area of the city have begun to expand into adjacent residential streets, formerly white-owned and occupied. Houses or flats may be rented indirectly via white nominees

Table 6.2   Attitudes of white and Indian households towards segregation and desegregation

Question:	Are you happy with the present situation where residential areas in Pietermaritzburg are segregated by race?					
Responses:	Indians	YES	56%	NO	38%	DON'T KNOW	6%
	Whites	YES	64%	NO	26%	DON'T KNOW	10%
Question:	Should ALL areas within Pietermaritzburg be opened up to all races?					
Responses:	Indians	YES	61%	NO	33%	DON'T KNOW	6%
	Whites	YES	20%	NO	69%	DON'T KNOW	11%
Question:	If there were no longer any restrictions on where you could live within Pietermaritzburg, would you consider moving into an area formerly occupied by another racial group?					
Responses:	Indians	YES	47%	NO	29%	DON'T KNOW	24%
	Whites	YES	33%	NO	58%	DON'T KNOW	9%

Notes
n = 310 (Indian Household Survey, 1987).
n = 290 (White Household Survey, 1988).
(*Source*: Department of Geography, University of Natal, Pietermaritzburg.)

who sign leases. An apparent loophole in the Closed Corporati
which allowed closed corporations to purchase land in white
despite a black majority share in the corporation has been explo
properties are also being acquired in the more exclusive formerly all-white
suburban areas, in anticipation of changes in the Group Areas Act. The
necessary secrecy which surrounds such transactions makes the quantification
of change very difficult.

While the political future is being debated Pietermaritzburg is changing
inexorably in many other ways. This is particularly noticeable in the central
commercial districts where the emergence of new shopping centres and office
complexes is matched by the proliferation of pavement activities on the periphery
of the CBD, and increasingly within those previously 'sacrosanct' precincts as
well. In the vicinity of the major bus and mini-bus taxi termini in particular,
Pietermaritzburg assumes the character of cities elsewhere in Africa. Shop-
keepers vie with hawkers to tap the crowds thronging the pavements, and the
jumble of colours, noise and smells is in sharp contrast to the recently pedestrian-
ized CBD. Since the declaration of a free trade area within Pietermaritzburg in
1986, there have been signs of a 're-occupation' of their former trading areas
by Indian businessmen. It is unlikely that this re-occupation will recreate the
distinctive Indian townscapes of the pre-Group Areas Act era, given the
structural changes that have occurred within the Indian business community in
the interim: changes that have seen the rapid decline of the small family-owned
and operated shops.

The post-group areas city: Pietermaritzburg tomorrow?

Planning for change

The City of Pietermaritzburg has recently attempted to grapple with its future
by implementing a strategic planning exercise, the Pietermaritzburg 2000 Project,
which encouraged wide public participation. The subsequent research has
brought some sobering statistics to light. The population of the metropolitan
area as a whole is likely to exceed 1.4 million by the year 2000 (i.e. more than
double the present population). In the metropolitan area about 75,000 houses
will have to be built by the turn of the century. At the same time it is estimated
that over 350,000 new jobs will need to be created in the same period. Clearly
new, lower cost, ways of creating employment and housing people will need to
be found. This is already leading to deregulation and the greater acceptance of
informal employment and housing of a standard more compatible with Pietermaritzburg's increasingly Third World character. In Pietermaritzburg the planning authorities have already adopted a more accommodationist stance, and
have become actively involved in making provision for pavement traders, for
example, and in planning for the integration of informal markets into the
townscape.

It is significant that the strategic planning exercise for the metropolitan area as a whole has been carried out by the white-controlled core city of Pieter-maritzburg. At present, physical planning of most of the black sectors of the city is carried out in Pretoria by State departments or agencies. The City Council of Pietermaritzburg is but one of 10 agencies responsible for the planning and management of what is a relatively small economically inter-dependent region.

It seems clear that reform, such as the repeal of the Group Areas Act and the Separate Amenities Act, which will 'deracialize' cities, will precede any more fundamental re-ordering of local government which may take place, or any policies aimed at a redistribution of resources rather than increasing access to them. Initial post-group areas changes in metropolitan Pietermaritzburg are therefore likely to occur within the unsatisfactory and fragmented framework of control which currently exists.

Post-apartheid Pietermaritzburg will have much more to contend with than redressing shortages and imbalances in the provision of housing and employment. Nearly a half century of institutionalized segregation has produced gross in-equalities in the distribution of urban amenities in general, ranging from education to recreational facilities, and the shadow of apartheid planning will be evident in the geography of the city for decades to come.

Dewar (1986) has asserted recently that one of the greatest challenges that will face post-apartheid South Africa is the reconstruction of the fragmented apartheid city, or 'closing the gaps of apartheid' (Collinge, 1990, p. 16). In Pietermaritzburg this fragmentation has resulted in the location of the urban poor at increasing distances from the city's commercial and employment heart, while paradoxically, empty, undeveloped or underdeveloped spaces exist nearer the city centre. Since the abolition of influx control, the Group Areas Act has formed the final 'legislative fence' preventing the movement into spaces pre-viously denied black residents.

Scenarios of post-group areas change in Pietermaritzburg (Wills, Haswell and Davies, 1987; Wills, 1990) have focused attention on the likelihood of a process of 'implosion' occurring – a turning inward of the city – as those who have been denied the opportunity to settle closer to the city centre move into the currently underdeveloped interstices of the apartheid city. In addition, and in the short term, there is no doubt that many people within the Pietermaritzburg metropolitan area will continue to seek sanctuary in the city from the political violence that has become endemic in the African residential areas on the city's edge. If formal housing and land suitable for development is not made available either nearer to the city centre, or to foci of employment, then the people may take the initiative and extend informal housing, currently concentrated pre-dominantly in peripheral collars around South African towns, nearer to the city centre. Currently undeveloped and man-made 'buffer strips', and under-utilized lots and buildings would be prime targets for such organized penetration of previously politically protected space. There are already precedents for this type of implosion elsewhere, for example the creation of Freedom Square shack settlement on a buffer strip, or racial no-mans-land, near Bloemfontein (Chapter 7).

Within Pietermaritzburg there is no better example of the contradictions of group areas planning than the Westgate/Slangspruit/Imbali complex (Figure 6.3). Here a belt of cleared land, deliberately treeless and fallow, acts as a buffer strip separating a white suburb of low-cost housing from the formal African township of Imbali, and the informal African settlement of Slangspruit. A new ring road, deliberately routed through the buffer strip, accentuates the divide. Slangspruit has developed at high density right up to the very edge of the buffer strip, occupied prior to the Group Areas Act by African landowners and market gardeners. There is an over-supply of houses in the predominantly blue-collar white suburb of Westgate, while in neighbouring Imbali, with a growing African white-collar population, there is a severe shortage of housing. The housing in Westgate in many ways represents a logical progression up the housing ladder for Imbali residents loath to invest their money on upgrading standard township houses, or reluctant to build new homes in a suburb severely under-provided with a range of amenities, and racked by political violence in

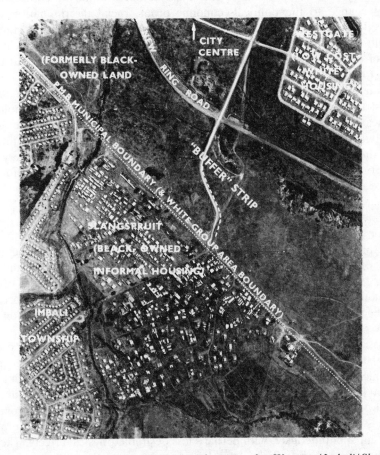

Figure 6.3 An example of apartheid city planning: the Westgate/Imbali/Slangspruit complex

recent years. In political terms, however, these geographical neighbours are worlds apart, with the white suburb a stronghold of conservatism.

The socio-geographical effectiveness of group area separation was well illustrated in this region of the city during 1990. At times violent confrontation between rival groups of residents of Imbali or Slangspruit took the form of pitched battles in the streets of those areas, often culminating in the intervention of the South African police or security forces. During these battles the white residents less than a kilometre away on their side of the buffer strip went about their daily lives without interruption, the only indication of civil unrest being groups of African women and children seeking temporary refuge during the day on the pavements, or in the gardens of sympathetic white households.

Suburban desegregation

Unless the negotiations between the government and the African National Congress and other elements of the 'mass democratic movement' result in a radical change to the political-economic order, racial discrimination will be replaced to greater degree by class discrimination in South African cities, and this will reduce the likelihood of rapid and major socio-demographic change in most of the currently white suburban areas of cities such as Pietermaritzburg, where for example, low incomes will inhibit the options of the bulk of the city's population in the early post-apartheid phase. Municipal estimates in 1986 suggest that only about 20 per cent of Indian and coloured households had incomes of more than R800 per month, while 80–90 per cent of the African households within the metropolitan area earn less than R450 per month. Thus while the pressures to relocate will be very strong indeed, only a very select minority will be able to do so, and the physical effects of political reform will thus be constrained by poverty.

The central city

Pre-reform desegregation has been proceeding most rapidly in the central cities, or older inner suburbs of South African cities (Pickard-Cambridge, 1988a, 1988b; Rule, 1988; Hart, 1989). In Pietermaritzburg the city centre is also likely to mirror political reform immediately. This is likely to occur not only because the inner city contains older, cheaper accommodation more readily accessible to the less affluent, but because there exists the potential to increase dramatically the resident population through infill and subdivision. Part of the Voortrekker legacy in central Pietermaritzburg are large, narrow lots which may be continuously developed along the street frontage, but which remain largely underdeveloped at the rear. There seems little doubt that, should the planning authorities not implement innovative and progressive schemes to encourage and control the 'infilling' of the city centre, the process will occur spontaneously, with perhaps less desirable physical effects. There are already signs of 'white flight' from the central area to the suburbs, due to the 'greying' process and pressures stemming from land-use change. In some sections of the city centre, long established white communities, complete with local schools and churches, may vigorously defend their racial exclusivity, while in other areas a supply of colonial red-brick houses regarded (by whites) as historically and architecturally attractive may survive change as exclusive 'gentrified' precincts.

Desegregation in public housing

Given the degree of central government intervention in the urbanization process in South Africa, public housing has played a key role in black urban communities. In Pietermaritzburg, public housing has always been segregated, with no possibility of mobility between, for example, the African township of Imbali and the large Indian council housing estate of Northdale. In terms of amenities, costs and location, African families in a post-group areas city may well regard the Indian or coloured housing estates as attractive alternatives, but the dire shortage of such housing is likely to constrain such transfers. Any lingering educational segregation is also likely to restrict initial movements, if the experience of other southern African cities is anything to go by (Wills, Haswell and Davies, 1987; Pickard-Cambridge, 1988b). There is relatively little white public housing in Pietermaritzburg, but the bulk of such housing is located near to the current African townships, and may well become an attractive alternative.

Conclusion

In essence Pietermaritzburg came into being as a racially divided town, and it has remained so divided until the present. Social segregation was progressively translated in an *ad hoc* manner into spatial segregation during the colonial years, and the 'neo-colonial' period preceding the coming to power of the National Party in 1948, which in turn resulted in the country-wide implementation of a comprehensive urban social design founded on racial separation.

Even the apartheid city which resulted has proved malleable, however, and the sheer momentum of urbanization has resulted in appreciable changes in Pietermaritzburg, reflected both in the townscape and the daily patterns of life in the city. Fundamental changes in the social geography of the capital of Natal await the dismantling of the remaining legislative pillars which have propped up an apartheid society, and its cities. It is debatable whether even the final scrapping of all vestiges of *de jure* racial discrimination will alone result in a more equitable urban environment in the short term. 'Closing the gaps' in Pietermaritzburg will take a long time.

7
BLOEMFONTEIN
D. S. Krige

Bloemfontein is one of the eight metropolitan areas in South Africa and has been the judicial capital of the country since 1910. Since the establishment of the Free State Boer Republic in 1854, Bloemfontein has been not only the administrative headquarters of the province, but also a stronghold of Afrikaner tradition. However, the economic centre of the province has shifted from Bloemfontein to the Orange Free State Goldfields since the 1950s. Furthermore, Bloemfontein is the core of the Bloemfontein-Botshabelo-Thaba Nchu region (BBT region) which is a proposed development axis of 70 kilometres and is perceived as a microcosm of apartheid planning with the following spatial components: Thaba Nchu district as an exclave of 'independent' Bophuthatswana; Botshabelo as an ethnic city for the Sotho, a catchment area for canalized urbanization and surplus blacks in the province, and a dormitory township; Bloemfontein as one of the ideal apartheid cities; the redrawing of bantustan boundaries (in an attempt to incorporate Botshabelo in the Qwaqwa bantustan which finally failed in 1990); three industrial development points (IDP); daily and weekly commuters and long-distance migrants; and the Bloem Area Regional Service Council (RSC) (Krige, 1988b, pp. 5–7).

The population growth of the three territorial structures of the region are shown in Table 7.1 for the period 1880–89. The absence of Indians is significant. As a result of an ordinance promulgated by the Free State Volksraad in 1890, all Indians were forbidden to settle or trade in the Free State. This provision was lifted almost a century later in 1986.

The colonial city (1846–1910)

Bloemfontein was founded in 1846 by Major Warden as a British outpost for the Cape Mounted Riflemen. Within a few years the military post had developed into a 'plattelandse dorp' which became the capital of the Boer Republic in

Table 7.1   Population growth in the BBT region (1880–1989)

Year	Bloemfontein				Thaba Nchu	Botshabelo
	White	African	Coloured	Total	Total	Total
1880	1,688	← 879→		2,567	13,452	–
1890	2,077	← 1,302→		3,379	18,496	–
1904	15,501	←18,383→		33,883	20,201	–
1911	14,720	10,475	1,730	26,925	25,251	–
1951	49,074	56,574	3,719	109,367	33,775	–
1980	90,900	115,420	14,760	221,080	56,602	60,000
1985	99,349	118,523	18,591	236,463	64,034	180,000
1989	108,000	140,000	22,000	270,000	70,000	220,000

(*Source*: Krige, 1990.)

1854. From the outset, Africans and coloureds were drawn to Bloemfontein as labourers. They were either accommodated on the premises of employers or in peripheral locations (Van Aswegen, 1968). Other than elsewhere in the Free State, separate residential areas were made available to the African and coloured residents at Kafferfontein and Waaihoek respectively (Figure 7.1(a)) (Schoeman, 1980, p. 35).

Bloemfontein experienced its first large-scale urbanization period between 1890 and 1904 when the number of residents in the town increased 10 times. The rapid population growth was chiefly caused by two events, namely the completion of the Cape-Johannesburg railway line via Bloemfontein and the putting into commission of the railway workshop, and secondly the Anglo-Boer War when thousands of labourers from white farms took refuge in Bloemfontein where a refugee camp had been set up. The rapid increase in the number of coloureds and Africans not only led to an intermingling in Waaihoek, but also to stricter measures by the City Council in respect of the entrance and sojourn of these population groups. A separate coloured residential area, Cape Stands, came into being in 1902 at the request of the Cape Boys (group of coloureds from the Cape Colony) who did not want to live in a mixed Waaihoek location.

The rigid application of the separate area principle to the various population groups in the colonial city phase of Bloemfontein made Davenport (1971, pp. 5–6) refer to the Free State as '. . . the most deliberately segregationist province of all' where a 'blueprint for absolute cultural segregation between the races in the towns' was established.

Bloemfontein as a segregated city (1911–50)

This period is characterized by two processes of ethnic separation, namely the outward displacement of Waaihoek's residents and the spatial separation of the African and coloured residential areas. The two processes form an integral unit, as the two population groups were re-accommodated in separate residential areas to the east of the railway line during the clearance of Waaihoek.

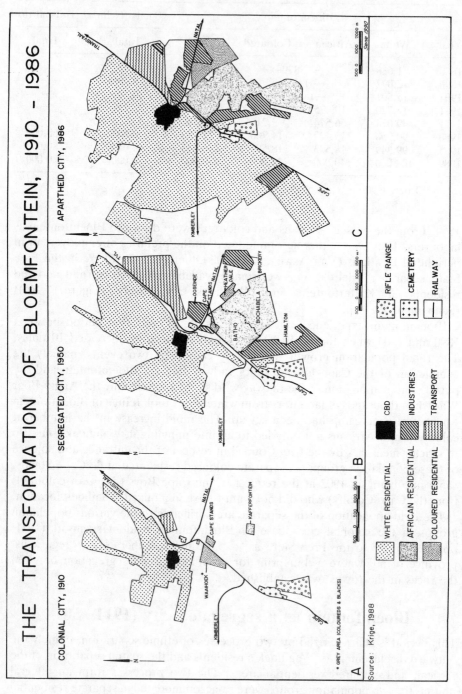

Source: Krige, 1988

Figure 7.1 The transformation of Bloemfontein, 1910–1986

Already in the Republican period the City Council was pressurized to move Waaihoek residents outwards because of the proximity of the area to a growing white sector. In 1917 the City Council decided to demolish Waaihoek and to resettle the coloured and African residents respectively in Heidedal (previously Heatherdale) and Batho, which are separated by the Dewetsdorp road (Figure 7.1(b)). The cleaning-up process started in 1919 with the removal of the railway compound. The unrest of 1925, which originated in Waaihoek, accelerated the cleaning-up process. However, the last of the more than 1,000 houses were only demolished in 1941 because of the cost of resettlement. With the demolition of Waaihoek, 80 years of an original part of Bloemfontein disappeared without trace. Today the area is occupied by industries and a white residential area.

The African residential area, Batho, was planned in 1918 and proclaimed in 1924 under the Natives (Urban Areas) Act of 1923. According to the proclamation, Kafferfontein and Cape Stands were included in Batho. The increase in African residents made it necessary to increase the land area of Batho. Land was added to it on the east by a number of proclamations from 1931 to 1949 forming Bochabela, the second of the four residential areas of the present African township of Mangaung (Figures 7.1(b) and 7.2).

In the 1930s Batho was referred to as the model African residential area. The reasons for this included the following: Batho is situated only 3.3 kilometres from the central business district (CBD); the provision of more advanced services like a pharmacy, the first crèche and post office in an African residential area in the country, a library, community and recreation hall, market, park with sports facilities and a bus service (Schoeman, 1980, pp. 285–6); and the well-known 'Bloemfontein System' which was a unique scheme for the erection of African housing in the country (Morris, 1981b, p. 28). This assisted building scheme was aimed at enabling prospective home owners to build houses. Building material and a number of approved building plans were made available at a low interest rate by the City Council. This resulted in home ownership by the majority of heads of families in Batho and Bochabela (Table 7.2) as well as a variety of building styles.

From the foregoing analysis it is clear that Bloemfontein already showed clear similarities to the apartheid model. The structure of the city is cast according to the sectoral plan whereby the distinctive living spaces of each population group can extend outwardly without any structural obstacles. The Cape–Transvaal railway line is a symbolic dividing line which divides the city into a white western and African eastern section. The railway line as buffer between white and African is further strengthened by the industrial areas, the premises of the transport services, shooting range and two cemeteries (Figure 7.1(b)). Whereas the colonial city phase was characterized by the residential separation of white and African, the second phase is distinguished by the residential separation of coloureds and Africans.

The apartheid city (1951–85)

As the segregated city of Bloemfontein already demonstrated strong similarities with the apartheid model, the idealized spatial patterns of the latter were achieved with smaller scale, less radical restructuring than in most South African cities (Figure 7.1(c)). Apartheid planning in the region was directed at the African population group especially, and the establishment of Botshabelo in 1979 was the climax of ideologically based spatial planning in the region.

In 1959 the Bloemfontein City Council applied for the proclamation of Heidedal as a coloured group area. This materialized four years later. In 1980, a second coloured group area was proclaimed at Rodenbeck (Figure 7.2) which has not yet been developed. According to the provisions of the Group Areas Act, the coloured residents of Mangaung, who lived mainly in the coloured ethnic island of Cape Stands, were compelled to move across the street to Heidedal, whereas the African residents of Heidedal had to move to Mangaung. However, the numbers actually moved were small. According to the 1951 census only 24 Africans lived in Heidedal, and the resettlement of coloureds involved no more than a hundred families. As a result of intermarriage and years of living together in Mangaung a number of coloureds remained in Mangaung living as Africans (Krige, 1988a, p. 86). The trade separation provision applied to only one coloured business, which was moved from the edge of the CBD across the railway line to Heidedal.

Mangaung was the area in Bloemfontein where apartheid planning had the greatest impact on the spatial patterns (Table 7.2). Figure 7.2 shows certain data about Mangaung that are related to Table 7.2. The establishment of buffer strips was the first change that came about. In 1953 the City Council was notified by the central government that no objections existed to further expansion of Mangaung, provided the new residential areas (Phahameng and Kagisanong) met the regulations in respect of buffer strips, namely an unoccupied strip of 180 metres separating the adjacent farms and industrial area, and one of 450 metres between Mangaung and the (coloured) Rodenbeck group area. As the buffer strips are part of the total area of Mangaung (1,792 hectares), a considerable part of the land area of Phahameng (35 per cent) and Kagisanong (25 per cent) was used for ideological friction-prevention zones.

The second change in spatial patterns was the zoning of the three dominant African groups, namely the Tswanas, South Sothos and Xhosas who represented 42 per cent, 34 per cent and 20 per cent of the African population respectively in 1980. The next change was the provision of houses by the central government to address the acute housing shortage. With the development of Kagisanong in the early 1960s, project 3,000 was launched whereby 3,000 houses were built. A total of 6,043 houses were built in Kagisanong before the provision of family housing was suspended in 1968.

The temporary status of African residents in Mangaung reached a climax in 1968 when all further land additions to the African residential area were frozen, and it was decided that family housing would be provided in the Thaba Nchu area in future. As a result the housing backlog accelerated alarmingly. In

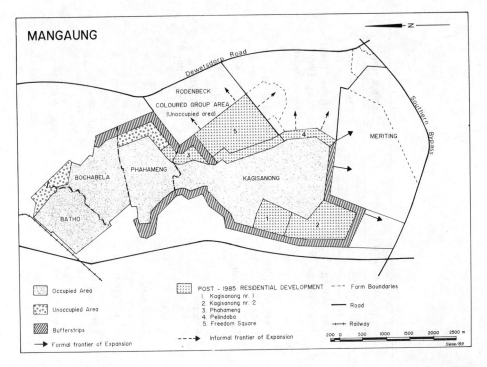

Figure 7.2   Mangaung

1975 already there was a demand for more than 6,000 housing units. Population density and over-occupation of existing structures and premises, together with the prohibition on home ownership, increased the unattractiveness of a once model residential area for the established residents as well as the prospective new residents.

The establishment of Botshabelo, 55 kilometres east of Bloemfontein and with an area of 10,000 hectares, dominated the changing spatial patterns of the region. Botshabelo was developed for two fundamental reasons, namely to accommodate the population growth of Mangaung, and to create a place of resettlement for the Sotho in the Thaba Nchu district after the 'independence' of Bophuthatswana. The development of South African cities can be interpreted on the basis of exclusion and creation processes (Krige, 1988a, pp. 103–108). The increasing pressure placed on the entry and residential rights of Africans in apartheid cities and the canalizing away of the urbanization process of this population group to nearby bantustans point to the exclusion process, while the ideologically created urban settlements in the bantustans represent the creation process. Morris (1981b, p. 144) refers to these processes as follows: 'It seemed that the overall intention was to make urban African townships as unattractive as possible, both to discourage further settlement and to encourage a reverse movement to the bantustans.'

From the outset Botshabelo was planned to develop into a fully-fledged and

Table 7.2 Changing spatial patterns in Mangaung

Period	Pre-apartheid pre-1953	Apartheid 1953–85	Neo-apartheid 1986–90
Suburb	Batho (1918) Bochabela (1931)	Phahameng (1956) Kagisanong (1964)	Pelindaba (1989) Meriting (1989) Freedom Square (1990)
Buffer strips	None	Vary in width from 180–450m	Relocation redevelopment[1]
Ethnic zoning	Inter ethnic	Ethnic zoning	Inter ethnic[1]
Schools	Inter ethnic	Ethnic	Inter ethnic[1]
Expansion of territory	Yes	Freezing	Yes: Meriting Freedom Square
Number of houses	4,000	11,000	6,500[2]
Type of housing	3,400 house ownership 600 rental	1,627 house ownership 6,043 rental 3,300 self-help	Elite[1] Upgrading[1] Self-help[1] Informal[1]
Home ownership	Yes	No	Yes[1]
Freehold	No	No	Yes (5,676)[1]
Single sex quarters	Compound of hostels	Hostel	Changing function

Squatting	No	Backyards	Backyards Open spaces + bufferstrips Freedom Square
Resettlement	From Waaihoek to Batho	Between Heidedal and Mangaung; from Mangaung to Botshabelo	Probably squatters on open spaces

Notes
[1] Applicable in all suburbs.
[2] This number of houses are predominantly in Kagisanong (1,800), Phahameng (400), Pelindaba (1,000) and Freedom Square (3,000) – (see Figure 7.2).

t city, which would not only have a residential function with
ig services, but which would also fulfil trade and industrial functions.
The city was planned formally, although its initial appearance was that of an
informal settlement as a result of the predominantly informal housing structures.
By 1989 the percentage of temporary structures had decreased to 54 per cent of
the total number of 37,000 housing units on 34,000 occupied stands as a result
of the upgrading of housing units (Krige, 1990).

Next to the national road from Bloemfontein to Maseru an élite residential
area with fully serviced stands, tarred streets, street lighting and élite housing
has been developed, which together with the industrial area is known as 'The
Curtain'. Behind 'The Curtain', poverty and unemployment are concealed.

Since 1979, Botshabelo has taken over the catchment function of 'surplus'
Africans in the Orange Free State from Thaba Nchu and Qwaqwa and also
serves as place of resettlement for the Sotho refugees from ethnically different
bantustans such as Thaba Nchu and Herschel, Transkei (Cobbett and Nakedi,
1988). In addition, Botshabelo is the catchment area of canalized African
urbanization away from the primary catchment areas in the province. According
to Bernstein the shift in the urbanization process leads to the shifting of the
Third World urbanization problems from the core areas to the periphery (cited
in South Africa, 1985, p. 161). Nevertheless the President's Council Committee
on Urbanization referred to Botshabelo as a model for Third World urbanization
where planning in advance enables the authority concerned to accommodate
rapid urbanization in an orderly way, which varies from informal residential
areas to high-level private homes (South Africa, 1985, p. 134). The catchment
function, whether as a result of resettlement or of canalized urbanization,
resulted in an increase of 180,000 residents from its establishment in 1979 until
1985 (Table 7.1), which makes Botshabelo one of the most rapidly growing
urban settlements for that period in South Africa.

As Botshabelo, like towns in bantustans, does not have a sufficient economic
basis, labour is the primary export product in the form of daily and weekly
commuters to Bloemfontein and as migratory labourers and monthly commuters
to the OFS Goldfields and other core areas. In 1987 it was found that 55 per
cent of the active labour force had jobs outside Botshabelo (Krige, 1988a,
p. 123). The number of daily commuters between Botshabelo and Bloemfontein
reached a climax of 14,500 in November 1988, but subsequently decreased to
10,600 in April 1990. The cost of a monthly ticket has also increased fivefold
from R12 in 1983 to R58 in 1989.

As Botshabelo serves as a distant extension of Mangaung and a dormitory
township of Bloemfontein, the latter has become a fragmented city, linked
together artificially with the help of a series of state subsidies. The transport
subsidy alone, whereby 60 per cent of the bus ticket of each commuter is
subsidized, amounted to R15.6m in 1989. In addition, the urbanization process
of the fragmented city is transferred to the periphery, but Botshabelo provides
part of the labour force of the core. This has delayed the growth of African
population in Bloemfontein, and the African percentage of the city's total
population actually decreased from 52 per cent in 1980 to 50 per cent in 1985.

In 1982 Botshabelo was proclaimed as the third industrial development point in the region, which is a clear proof of fragmented apartheid planning. By the standards of South Africa's overall programme of industrial decentralization, Botshabelo has been a modest success for the state. In 6 years 57 factories were in production employing up to 10,000 local workers. As one of the many designated 'growth points' Botshabelo receives state incentives in an attempt to attract local and international investment to the area. The level of these incentives is such that Botshabelo, and other 'growth points' can claim to be amongst the cheapest places from which companies can produce. Its financial cost to the state is high indeed, but it is the exploited workers of Botshabelo who bear the real burdens. Some of the incentives available are as follows: the state offers, *inter alia*, relocation expenses of up to R600,000; subsidized capital loans and rentals; subsidized electricity; housing for 'key' personnel; tender preferences; railage rebate; training subsidy; but most importantly, the state will pay 95 per cent of the wage bill tax-free, in cash, up to a maximum of R100 per worker per month for the first seven years. With the last-mentioned subsidy available, it is amazing that factory workers are exploited by being paid as little as R2 per day and that the average income per month was between R60 and R70 in 1987 (Cobbett, 1987). Such cheap labour has also attracted investment from South Africa's international friends, primarily from Taiwan, in an attempt to beat sanctions. As a result of the meagre salaries, 93 per cent of all work opportunities are taken by women, who are 'traditionally' lower paid.

The fragmented nature of the spatial development of the BBT region is the result of apartheid planning, during which dividing walls were erected between the First World and Third World communities. These social-engineering policies of the apartheid state are justified by the architects of apartheid by referring to the region as a development axis in which 'bundled development' takes place. Part of the 'bundled development' is the proclamation of Bloemdustria, 20 kilometres east of Bloemfontein (Figure 7.3), by the state in 1988 as an industrial deconcentration point. Bloemdustria, like Botshabelo, is part of Bloemfontein, and the latter's industrial development is being transferred to Bloemdustria as Bloemfontein's African population growth has been transferred to Botshabelo. In 1990 eight factories were in production in this industrial area. Two of these industrialists are from Botshabelo – an indication of the nomadic nature of industrialists in IDPs.

Bloemfontein as a neo-apartheid city (1986–90)

The effects of recent reform measures can already be seen in the spatial patterns of the BBT region. One of the most important changes was the abolition of influx control which means that canalized urbanization to Botshabelo will in due course be replaced by natural urbanization in Bloemfontein. Population growth in Botshabelo has decreased drastically since 1985 and thousands of serviced sites that were to have accommodated the 'orderly urbanization process' are still vacant. By the end of 1989, 400 permanent houses were vacant, which may indicate an outflow of people and/or the unaffordability of formal housing for the low-income group. It appears that the catchment function of Botshabelo has been changed to a transition function as labourers settle closer to their place of work to save travelling costs and time.

On the other hand, the population pressure on Mangaung and on the smallholdings around Bloemfontein has increased, which has led to squatting in both areas. One of the objectives of creating Botshabelo was to clear up squatting in Bloemfontein and Mangaung, but now squatting is already back as a permanent phenomenon. The African population on the smallholdings doubled to 26,000 from 1985 to 1989. In certain parts of Bloemspruit (adjacent to the eastern boundary of Bloemfontein) squatting has increased to such an extent that it has led to white flight and accompanying greying of the area.

Backyard squatting and the accommodation of lodgers in houses are common occurrences in Mangaung. In 1989 squatting also started in the buffer strips which led to the relaxation of the regulations regarding the use of these zones. The area known as Pelindaba was formalized and rudimentary services were supplied. A second informal occupation of land started in February 1990 when families looking for homes in Mangaung were mobilized in an organized way by the Mangaung Civic Association (MCA) to settle on the adjacent undeveloped Rodenbeck group area as part of the repossession of the land (Collinge, 1990, p. 16). Plots were measured out in an informal manner by the committee members of the MCA themselves, and they also handled the allocation of stands. The area was called Freedom Square (Figure 7.2) as this 'semi-liberated zone was won by the people .... which is perceived as a tremendous stride forward for the landless' (Collinge, 1990, pp. 16–17). This was where the Civic Association could meet the requirements of the poorer, homeless, landless residents of Mangaung – something which neither the City Council of Mangaung nor the private developers could do. The momentum of the land invasion process is illustrated by the approximately 3,000 structures that were erected within three months. The MCA is also looking towards ways of transforming Freedom Square from a village of zinc to one built in bricks and mortar. The illegal occupation of land and the problems of service provision led to discussions between the provincial authorities and MCA. It was decided to formalize Freedom Square; the Province would provide dirt roads and emergency services until the area could be supplied with rudimentary services by the City Council of Mangaung at a later stage.

Freedom Square has certain important consequences for the future spatial development of the region. The dynamics of the informal occupation of adjacent government land have resulted in sites being available to the poor who are normally provided with premises in Botshabelo only. How payment of services and of already occupied stands will be made is not yet clear. The expectation is that the government will switch to a system of subsidized stands which will mean that the residents of Freedom Square will pay a minimal amount. In addition it is assumed that informal housing on the frontier of expansion will be a permanent phenomenon. Although an eastern boundary has been established for occupation, the successful occupation of Freedom Square can convert the eastern boundary into an eastern frontier of expansion which could be occupied by squatters on the smallholdings and by residents of Botshabelo. This frontier of expansion in the direction of the Bloemspruit smallholdings may lead to a gradual process of greying of the smallholdings.

Squatting also took place in Mangaung itself on open spaces, buffer strips and even school premises. More than half of the approximately 700 squatter structures on open spaces are in Tambo Square, an area zoned for a city centre and only 3 kilometres from the Bloemfontein CBD. In reaction to the establishment of squatter structures in and around Mangaung a local newspaper with right-wing political sentiments referred to the squatter camps as a breeding ground for unrest where people's courts and comrades are present in the liberated areas, and said that white safety is threatened by this (*Bloemnuus*, 4 May 1990).

It appears that the lifting of influx control had a clear influence on the choice of settlement in the region, and that the unbanning of the ANC has a visible impact on the settlement patterns, for instance the shack sprawl on the outskirts of the city as part of the repossession process of the land.

Because of the unfreezing of the provision of family housing and of the territory of Mangaung, formal development is taking place in a southerly direction (Figure 7.2). The first élite residential areas with fully serviced stands and tarred roads were developed for the high-income (80 per cent) and middle-income (20 per cent) groups at Kagisanong, Extension 1. The cost of the 1,200 stands varied from R18 to R25/m^2 (as against the 25 c/m^2 for similar stands in Botshabelo). The last section of the original Mangaung, Kagisanong, Extension 2, is being developed and 3,320 stands for the high-income (60 per cent) and middle-income (40 per cent) groups are planned.

Blomanda Housing Company has purchased two farms for township development south of Mangaung. The area, known as Meriting, was incorporated with Mangaung in 1990 and approximately 14,000 stands are envisaged on the 771 hectares. According to the developers it is planned to divide the number of stands equally among the high, middle and low-income groups.

It appears that the availability of land, whether by informal land invasion or formal development, should not be a stumbling block for future African urbanization. Factors that may be restrictive are the provision of affordable premises for the poor and the handling of the process by the City Council of Mangaung, which does not have the necessary financial resources to handle the influx of people effectively.

Further changes in Mangaung are the abolition of ethnic zoning amongst Africans, the restoration of home ownership and the introduction of full property rights. As a result of the long tradition of home ownership in Bloemfontein and because 34 per cent of the rented houses have already been sold, 77 per cent of all African houses are in private possession. In addition 31 per cent of all stands were registered as leaseholds (which could be changed to full property rights later) by the end of April 1990. The relatively high percentage of leasehold registrations reflects the selling of rented houses (land is registered at the same time), the development of new stands and the upgrading of existing structures.

As a result of the establishment of the Bloem Area RSC which became operative on 1 May 1987, R12.5 million was made available to the Mangaung City Council for development projects in the next three years. The amount represents 53 per cent of total RSC expenditure over the period to the five participating authorities (the others are three white and one coloured local authority). With the aid of these funds 68 per cent, 42 per cent and 27 per cent of all houses have water, flush toilets and electricity respectively. The funding by the RSC is of cardinal importance as only 53 per cent of all rent and services charges were collected for March 1990, which indicates a silent boycott. The total rent and services charges in arrears amounted to R6.8 million by the end of March 1990.

To the west of the railway line a section of the central business area has become a free trade area, the 'white by night' regulation (quarters for domestic workers not allowed in certain white suburbs) has been lifted in all residential areas and Indians may settle and trade in the city. However, since their arrival in Bloemfontein, the Indians have been living in a number of prescribed blocks of flats in the CBD according to a permit system, whilst the political debate over finding a suitable group area continues. Meanwhile the City Council, whose members all belong to the National Party, approved the opening of all facilities in principle at the end of 1989, but a further investigation into the possible implications will be carried out. It is expected that the greatest resistance will be in respect of the public swimming baths, bus service, public halls and library (*Die Volksblad*, 9 May 1990).

Bloemfontein in the 1990s

Finally the scenario for the expected spatial patterns of the BBT region for the year 2000 (Figure 7.3) will be described. It is assumed that the Group Areas Act has been repealed, that all forms of desegregation followed residential desegregation, and that the bantustans have been incorporated in a united South Africa.

It can be expected that no major changes will take place in the city structure of Bloemfontein in the short term. Although illegal greying is already taking place on the smallholdings in the Rodenbeck area, the potential for greying by invasion and succession to the west of the railway line within the next 10 years is considerably lower for several reasons. The present city structure is the product of 140 years of ethnic separation planning with the Cape–Transvaal railway line as the separation mechanism between white and black. The pre-apartheid Bloemfontein was cast according to social apartheid principles that have been refined since 1950 to conform to specific political objectives. Bloemfontein is likely to remain a predominantly social apartheid city in the face of such a long-established racial ecology. The coloured and African residents have not been disrupted by resettlement programmes for the past 40 years

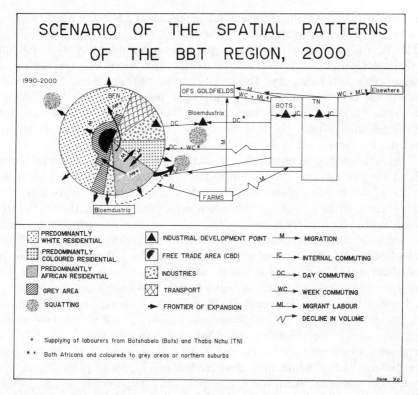

Figure 7.3   Scenario of the spatial patterns of the BBT region in the year 2000

(apart from those who had to move to Botshabelo, and from Cape Stands), and the residents have a long association with Heidedal and Mangaung respectively. Hart (1989, p. 86) asks whether all South African cities follow the pattern of ghettoization like some areas of Johannesburg, and rightly argues that 'it is likely that cities with Afrikaans cultural antecedents such as Pretoria and Bloemfontein might be expected to respond differently to African invasion than those cities with a more liberal tradition such as Cape Town and Johannesburg.'

Morphologically the city is cast according to the sectoral structure, with Heidedal and Mangaung both in a favourable position in respect of the CBD and industries. This geographical proximity to the heart of the city and job opportunities may decrease the greying potential of the white sector. Available space adjacent to Mangaung, as well as the sectoral structure of the city, facilitates both formal and informal extension and development for Africans in a southerly and easterly direction. Consequently, there is sufficient extension potential to accommodate the African urbanization process effectively and meet the different housing needs of all income groups at the same time. The upgrading of services and facilities in Mangaung enjoys high priority at present to increase the general quality of life in the township. It can be assumed that

ractive the living space in Mangaung, the less the need to migrate
juality residential area would be.

The social separation of the different population groups of the city is embedded
in the historical-political approach of spatial separation, and the railway line
represents a social boundary. The social distance between white and African is
greater than between white and coloured and between coloured and African.
The probability is therefore that more coloureds than Africans may settle
across the railway line, but a tendency for Africans to settle in an enlarged
Heidedal can be expected.

Economic realities reinforce this prediction. Most households on the eastern
side of the railway line cannot afford the cheapest houses owned by whites.
However, the black middle class have the financial ability to penetrate the
lower-income white suburbs, which are mainly situated along the north–south
railway line. It is precisely in these areas, however, that the strongest opposition
to residential integration exists as a result of right-wing political sentiments.
Individual black families in the high-income group have the ability to settle to a
limited extent in the wealthy northerly suburbs. Currently, however, there is a
shortage rather than a surplus of white housing, although not on the scale of
the housing backlog in Heidedal (1,000 units) and Mangaung (8,000 units).

It is expected that the most affluent black families will be the first to move
over the railway line. Middle-income families may rent flats in the CBD and
fringe area, which may lead to greying and accompanying white flight. Hilton is
regarded as the first white suburb where greying may take place, because the
residents of this area, like those of the CBD, lack neighbourhood cohesion as a
result of the increasing number of light industries in the area. It is also
suspected that coloureds (and Indians, because a separate residential area has
not been proclaimed for them) will settle more readily across the railway line
than Africans given the smaller social distance, and also because designation of
Rodenbeck as a coloured group area has been negated by the informal occupation
of the area by families from Mangaung. In addition, the natural hinterland of
Heidedal is characterized by slum conditions where squatting has led to the
withdrawal of whites and decreasing property values; in 1990 only R100 was
offered for one 5 hectares smallholding, for example (*Die Volksblad*, 19 March
1990).

In the longer term the Bloemspruit smallholdings can be seen as the informal
frontier of expansion. The area is not only favourably situated in respect of the
CBD and Bloemdustria, but basic infrastructures such as water, sewerage,
electricity and roads already exist, which may accelerate the process of occupation. In the open space between the Hamilton industrial area and the
railway line (Figures 7.1(c) and 7.3) the first new non-racial residential area
seems likely to develop.

In the high-income areas of Bloemfontein, the growing Third World character
of the city will probably lead to increased privatization of schools, health and
other facilities, as well as the extension of the neighbourhood watch system.

The overall picture is thus one of limited residential integration in the short
term. In this respect Bloemfontein may well be representative of many smaller

urban areas in South Africa, especially those with a predominantly Afrikaans population. In the wider BBT region, the establishment of Botshabelo was the greatest single disruptive factor, and will be the major apartheid inheritance. As a result of the changing political and economic climate in South Africa, its artificial boom in population growth and industrial development has already collapsed, which portends an even bleaker future for the remaining inhabitants of this dormitory settlement in the post-apartheid era.

Acknowledgements

An earlier version of this paper was prepared for an article in the *South African Geographer*. Mrs G. C. Jankowitz is thanked for the preparation of the diagrams.

8
KIMBERLEY
G. H. Pirie

Kimberley is best known for diamond mining and for the glittering as well as the flawed personalities and exploitative labour arrangements that crystallized in its harsh environment (Roberts, 1984; Turrell, 1987; Worger, 1987). Yet the roistering mining camp which was established on featureless terrain between 1879 and 1881 also occupies an important place in South African urban affairs: it was here that the first formal strategy of racial residential segregation was devised and implemented.

When corporate, deep-level industrial mining knocked aside the independent diggers on the diamond fields in the 1880s, haphazard settlement in the cosmopolitan tent town gave way to a more rigidly supervised and segregated pattern of housing. In order to disunite black and white workers at a time of labour unrest, white mine employees were settled in an isolated company village. The segregation of black males into barrack-like compounds fulfilled the same objective. In addition, the practice also allowed the monopolistic De Beers Company to stem labour desertion, to foster tighter discipline, to overcome fluctuations in the supply of migrant labour, and to reduce the theft of diamonds (Mabin, 1986).

Self-employed (and unemployed) black and white people in early Kimberley lived at scattered points throughout the town, sometimes together, as in the unplanned Malay Camp (Figure 8.1). This district was initially settled by Malay transport riders in the 1860s. In time the Camp was home to a varied population of Africans, Indians and some whites. By 1892 seven fairly substantial peripheral 'locations' had also emerged. These housed approximately 8,000 African people, only slightly fewer than the number in the mine compounds. Four of the sites were known simply as No. 1, 2, 3 and 4 locations; the others were named Bultfontein, Mankoroane's and Greenpoint. By the outbreak of the First World War the informal, privately-owned tenements and sacking- or packing-case huts which characterized the locations were being described as the most

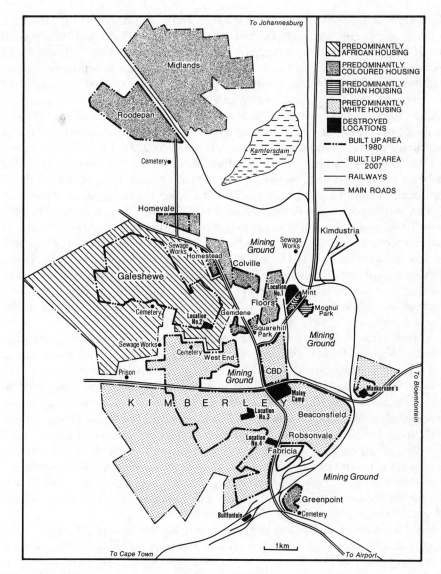

Figure 8.1   The changing racial geography of Kimberley

insanitary and disorderly in the Cape, and proposals were made to start public
housing for Africans.

The segregated city (1920–50)

The passage of the 1920 Housing Act gave the Kimberley Town Council access
to central government funds and enabled it to begin its first housing programme
for Africans. Sites for 350 houses were selected west of No. 2 location, and 120

plots were set aside at Greenpoint, adjacent to the old cemetery. By the time building had begun, the Natives (Urban Areas) Act of 1923 had been passed and it became possible to segregate Africans forcefully. The Act was made applicable in Kimberley in the following year. House construction in the two locations was not undertaken with any great urgency: by 1930 275 units had been built in No. 2 location, and 95 in Greenpoint[1].

The extension and formalization of two locations in Kimberley was followed by the consolidation of African settlement. Not only did the number of African men living in mine compounds in Kimberley fall to 2,500 by 1938, but by the late 1930s No. 4 location, as well as Bultfontein and Mankoroane's locations had disappeared. Shortly, by 1941, No. 1 location had also been eliminated and the Ashburnham industrial area was established on the ruins. In the remaining four locations, population densities rose. In No. 2 location 7,500 people crowded onto 1,400 plots; the number of lodgers was estimated at 2,000. In No. 3 location some 400 people were housed on a mere 31 sites. Almost 4,000 people were accommodated on 480 stands at Greenpoint. Conditions were atrocious, with most Africans living miserably in 'badly built, ill-ventilated hovels'. The 2,000 or so Africans who sheltered in backyard rooms provided by their white commercial, domestic or industrial employers, or who rented shacks in the Malay Camp, would have been housed little better.

A great many coloured people in Kimberley lived in equally unpleasant circumstances in the Malay Camp. Others crammed into squalid properties in racially mixed Beaconsfield which, like nearby Greenpoint, lay adjacent to the railway line and railway yards. Two hundred coloured people also lived in No. 3 location. This, the smallest location in Kimberley, contributed to the 'indescribably depressing' melange of black housing. The location was grossly overcrowded, and the inadequate sanitary arrangements aroused concern among the civic authorities because of the location's proximity to the city's waterworks. In view of the perceived public danger, No. 3 was soon earmarked for demolition. It was not until 1944, however, that the inhabitants were uprooted to No. 2 and Greenpoint locations. Compensation was paid to home-owners, and demolition allegedly occurred without the slightest complaint.

Residential development and improvement in Kimberley was severely stunted for the first 70 years of the town's existence because all land was held in leasehold from the De Beers mining company. This arrangement altered in 1939 when the company donated the major part of its estate that fell within the municipal boundary to the Town Council. The deed of donation was agreed to by the company, the Council and the Provincial Administrator; no other parties were consulted. The deed incorporated two conditions that were to have a marked effect on the racial geography of Kimberley. One clause stipulated that the Malay Camp would have to be abolished and that all land leases there would be terminated at the end of 1953. A second clause was intended to prevent black people returning to the improved Malay Camp site, and to prevent other concentrations of black people emerging spontaneously elsewhere in the city: except in certain places, Africans, coloureds and Indians could not buy, occupy or use donated land.

Restricting land occupancy in large parts of Kimberley to whites was tantamount to enforced segregation conducted at the behest of, or at least with the connivance of, the local authority. It aroused earnest complaint from the Coloured Ratepayers', Standholders' and Citizens' Association which observed that coloured people had seldom expressed any wish to encroach into predominantly white neighbourhoods. Much of the logic behind the exclusionary clause fell away when the Council decided to transform the Malay Camp into a complex of civic buildings rather than into a new residential area, but the blatant racial discrimination still rankled. Materially, the clause made it impossible for black people to profit from dealing in property or by letting accommodation.

While the debate raged about the legality and fairness of the deed of donation, in the early years of the Second World War the Town Council began constructing 80 sub-economic houses for coloureds who worked in the ammunition factory which was previously a mint. At the far north-east edge of town, across a railway track, near the city abattoir and sewage disposal works, Mint township was the first of several public housing ventures for coloureds in the same part of Kimberley. The next scheme, at Floors, was larger and was designed to absorb people who were evicted from the Malay Camp. The City Council commenced construction of 500 houses there in the late 1940s.

The removal of black people from the Malay Camp after 1953 was assured as early as 1939, but the legislative opportunity to destroy the congested, ramshackle quarter only arose when the 10-year-old Slums Act was made applicable to Kimberley in 1944. By then some 3,000 people (the vast majority of whom were coloured) jammed into a mere 400 houses that were jumbled around Bombay, Canton, Ceylon, China and Orient streets. Over a period of almost a decade, three-quarters of the dwellings in the Camp were obliterated. By mid-1953 the number of residents had declined to 755. A handful of whites, Indians and Chinese remained together with 441 coloureds and 272 Africans.

The establishment of the Floors and Mint townships breached but did not eliminate neighbourhoods where coloureds and whites rubbed shoulders in Kimberley in the 1940s. The creation of a housing scheme for coloureds at Robsonvale would have taken segregation a step further had the Council not diverted 120 of the 144 houses to white soldiers returning at the end of the War in 1945. As it was, this re-allocation contributed more to the perpetuation of residential mixing than to its demise. Despite efforts to concentrate coloured homes in demarcated districts, in the early 1950s only a quarter of Kimberley's coloured population of some 12,500 persons lived in the three public housing schemes. The remainder occupied private accommodation chiefly in West End and Beaconsfield. Gradually, in the 1950s, racial mixing waned as the Town Council developed new housing estates for coloureds at Colville, Gemdene, Homestead, Homevale and Squarehill Park. These contiguous townships form a distinctive wedge of coloured housing to the north of the city centre.

The apartheid city (1951–90)

The racial residential segregation that had first been engineered locally in Kimberley in the 1940s by De Beers and the Town Council was reinforced by central government intervention within a decade. The Group Areas Act which made racial prohibition statutory took effect in Kimberley in 1951 after which land transfers between people of different race were forbidden. Later, as part of the programme of phased racial zoning, land ownership was restricted along racial lines. Subsequently it was stipulated that the occupier of a house must be of the same race as the owner. In time, racial segregation was extended to retailing. When Floors was declared an exclusively coloured group area, Indian traders were required to vacate trading sites there. The multi-racial Kimberley Civic League and the Non-European Unity Movement arranged a boycott of trade site sales, but this opposition was only token.

The Town Council, which had for years been engaged in its own socio-spatial manipulations, co-operated with the government's efforts by undertaking a detailed survey of Kimberley as a preliminary to race zoning. There followed a lengthy round of proposals, criticisms and revisions. At first the Kimberley Civic League and the Non-European Unity Movement refused to participate in these exercises. Their stance mellowed when it became clear that race zoning would go ahead without them, and that even a modicum of participation might secure some benefits for the victims of Group Areas legislation (Mather, 1985).

By the mid-1950s consultation and planning for race zoning was nearing an end. It was not until four years later in 1959 that plans were finalized. The massive population dislocation which was envisaged affected race groups un-equally. A mere 300 whites, less than two per cent of the total, were affected. Many resided in Homestead, a racially mixed commonage outside the municipal boundary which was flagged for coloured occupation. The entire Indian and Chinese population of Kimberley (1,450) was affected by the group area proposals. For the first time this small group of people (mostly traders whose homes and shops were dotted all over the city) were obliged to congregate together, and away from whites. Land was set aside for this purpose first at Mint township from which coloureds were evicted, and then at adjacent Moghul Park.

Kimberley's coloured population was hit hardest by prescribed race zoning. Between 9,000 and 10,000 people (half the coloured population), were relocated and rehoused in Colville, Floors, Gemdene, Homestead and Homevale. The disruption involved more than just loss of homes. Coloureds lost many of the facilities which they enjoyed in central Kimberley: the sole Teacher's Training College and Coloured High School; eight of the 11 primary schools; a community centre; practically all the churches and several mosques; two cinemas, tennis courts, playing fields on the old site of No. 3 location, and a swimming pool (South African Institute of Race Relations, 1959).

By the end of 1962, 402 families had been resettled in line with group area plans in Kimberley. At that stage, 90 per cent of the 6,400 families in the city were correctly located in a designated group area: a further 677 families still had to be relocated.

Although the Group Areas Act was not targeted at Africans, those living in Greenpoint were affected indirectly in as much as the township was initially zoned 'white' and its African occupants were obliged to relocate to Galeshewe. After the removal of No. 1 and No. 3 locations, Africans who did not live on their employers' premises could only reside at Galeshewe or Greenpoint. By the mid-1950s Galeshewe had emerged as the principal concentration of Africans. Some 30,000 people lived there as opposed to 10,000 in Greenpoint where 300 coloureds, 18 Indians and 12 Chinese also resided.

Greenpoint was preferred by Africans who worked in the Fabricia industrial township, in the adjacent railway yards and workshops, and as domestic servants in nearby white suburbs, but the government was adamant that Africans should be confined to Galeshewe. Undivided township administration mattered more to apartheid planners than the personal inconvenience and expense of a longer journey to work for Africans. The Town Council pressed to retain housing at Greenpoint but the government had its way when it threatened that unless a start was made on removals from Greenpoint there would be no more funds forthcoming for the development of Galeshewe. In the event, a vigorous programme of construction there contributed 850 new houses and one migrant worker hostel by 1962. Zones of housing were differentiated along ethno-linguistic lines, with sections being earmarked for the Tswana (90 per cent), the south Sotho, and others. The demolition of Greenpoint began in April 1963 and by the end of the year more than 550 families had been moved across town to Galeshewe. Among the residents who were strongly opposed to forced relocation were an estimated 700 people who considered themselves members of the Griqua tribe and who urged that Greenpoint be dedicated for their use.

The pattern of residential apartheid that emerged in Kimberley in the 1960s formed the spatial matrix for residential development in the following decades. Provision was made for white population increases by proclaiming new suburbs to the south and west of the city centre. In the 1970s when the African population reached 50,000 (47 per cent of the city total), Galeshewe was extended to the north and west. For the first time the dull uniformity of public housing was relieved by the appearance of the first employer-assisted housing schemes such as the one in north-west Galeshewe which was begun by De Beers in 1973.

In the early 1970s Kimberley's coloured population of 25,000 was fast beginning to equal the 30,000 strong white population. The existing belt of coloured housing might have been stretched north-east of Floors but, perhaps because of the presence of the sewage works, this land was set aside for the Kimdustria industrial township instead. In turn, the more remote site at Roodepan which had initially been earmarked for industrial development was switched to residential use. Between 1974 and 1979 coloured families began occupying public housing at Roodepan, 14 kilometres north of the city centre. The hasty erection of inexpensive structures on poor clay ground had its price: walls began cracking and repairs were necessary to half the houses.

As a climax to the progressive racial segregation of residential space in Kimberley, in 1978 Galeshewe achieved the status of a self-governing town. Responsibility for the management, financing and planning of the township shifted from the Kimberley Town Council to an elected community council. This new creature of apartheid was doomed from the start by its shaky tax base and lack of political support. Over the years Galeshewe residents became disillusioned by the inability of their Council to improve conditions (which were at their worst in old No. 2 location) and to overcome the housing backlog: there was a shortage of more than 2,300 homes in 1987. By then approximately 67,000 people lived in the township, 1,400 in hostel accommodation and 3,200 in backyard shacks. An increasing number of Africans lived in new housing built by private contractors in the well-serviced peripheral suburbs of Galeshewe such as Ipeleng, Redirele and Retswelele. Those who could not escape the inferior conditions in the older sections of the township threw their weight behind the Galeshewe Civic Association and the Galeshewe Tenants Association which recently capitalized on massive rent arrears by declaring a formal rent boycott.

In contrast with the situation in Galeshewe, the coloured and Indian areas of Kimberley continued to be managed by the Kimberley City Council via advisory bodies known as the Coloured and the Indian Management Committees. The direct link to the City Council purse was still not enough to stem severe housing shortages. In the late 1980s the waiting list was variously estimated to include between 3,500 and 5,000 households.

By 1980 apartheid had reached an advanced stage in Kimberley in the residential, retail and public spheres. Enforced segregation of homes had been complemented by, among others, segregation of health care, displacement of the Indian retail market away from the centrally located City Hall to an eccentrically situated bazaar, and dislocation of the black bus and taxi termini. Then, after two decades of rigidity, apartheid finally began to show signs of fraying in Kimberley. In 1986 the government approved the declaration of practically the entire central city of Kimberley as a Free Trade Area, and the City Hall was opened for use by all residents. A public swimming pool which had previously been reserved for whites was desegregated in 1990. Residential apartheid was even relaxed occasionally in the late 1980s. One health care organization and two coloured families were permitted to locate in a designated white area temporarily until they could find alternative accommodation. Concessions of this kind were not always overlooked by whites. Between July 1989 and March 1990 12 complaints were investigated by the officer appointed by the government to deal with group area infringements.

The survival of Greenpoint as a site for even a small number of coloured homes (approximately 300) is the most glaring failure of the programme of residential apartheid in Kimberley. Through decades of neglect spawned by uncertainty the physical fabric of the township has deteriorated. Recently, a local Action Committee organized a rent boycott and petitioned the Town Council about the lack of water and sanitation facilities. Now, in 1990, three years before the lease of Greenpoint from De Beers expires (the land was not part of the 1939 donation), the township's residents are imploring the Town Council to upgrade the township and formally proclaim the 2,000 available sites for coloured housing.

The post-apartheid city

The concrete legacy of a century of racially motivated housing and infrastructure provision in Kimberley is one of the key considerations which will shape the future pattern of the city. The shortage and unaffordability of land and houses suggest that post-apartheid Kimberley will not be characterized by rampant residential desegregation. In addition, the dominance of Afrikaners among the city's white population (approximately 70 per cent) gives Kimberley conservative leanings: whites will not depart from their homes and suburbs easily, and blacks will not find the suburban property market wide open. Certainly, the town planning scheme which was launched in 1987 for a 20-year horizon was firmly rooted in the race consciousness of a traditionally conservative white City Council, and is not aimed at overturning the inherited patterns of territorial apartheid nor at dissolving the congruent geography of social ties.

Official estimates are that by 2007 there will be 100,000 more people living in Kimberley than in 1990. The population is forecast to be 267,000, including 133,000 Africans, 82,000 coloureds, and 51,000 whites. The projected increase is likely to be absorbed increasingly in the commercial and service sector where 60 per cent of economically active people are currently employed, and in industry which is eligible for government decentralization incentives. The location of homes for the increased population will be dictated in part by the restricted directions of city growth. Eastward expansion of the city is limited by mining land and the Kimdustria factory belt. To the south the city is hemmed in by the Fabricia industrial area, the airport and the railway yards, all of which present a noisy and polluted environment. With these geographical realities in mind, the projected absolute growth of the white population is currently being catered for by plans to increase residential density and land availabililty in the south and western sectors which are still designated 'white', and to eliminate the overprovision of open space. Extensions are also likely to De Beers white housing estate on the eastern fringes of the city along the Bloemfontein road.

The land requirements for future white housing, shopping, schooling and recreation are slight compared with those which will be needed to accommodate coloured and African population growth. The limited possibilities for increasing residential densities and utilizing open space more effectively in coloured and African townships means that by 2007 the land currently available for coloured housing must be increased by a factor of 1.6, and that for Africans doubled. Space which has been set aside for future coloured housing includes land to the south and west of Homevale and, more important, a vast swathe in the Midlands area which is intended for 10,000 sub-economic houses. Geotechnical tests have been conducted to assure prospective residents that their houses will not suffer the same fate as those in neighbouring Roodepan. The planned development of coloured housing at Midlands implies a long and expensive journey to work, but the costs and road traffic implications might be eased by the fact that the new settlement will lie astride a railway which could be upgraded for commuter transport.

Current plans for future low-cost public housing for Africans focus on additional extensions to Galeshewe. Corporate and self-build housing schemes for Africans will continue to flourish on the western and southern margins of the township. As a fresh departure, new middle-class and élite African housing is likely to invade the buffer zone which separates Galeshewe from white residential areas to the south. Meanwhile it is probable that Kimberley's first small non-racial Free Settlement Area will be declared here. In time the incursion will be much larger and will reinstate West End as a zone of racially mixed housing. At no time have the blocks of flats in and around the central area been considered as a potential 'grey area'.

Even when race ceases to matter in South Africa – when racial categories are suppressed and racial restrictions are removed from housing – vast tracts of new land will have to be opened for residential development. In a post-apartheid society there are sound reasons for avoiding the perpetuation of remote, sprawling, monochrome ghettos in the Galeshewe and Roodepan/Midlands areas. Wherever new townships emerge, however, future demography means that these will be predominantly black. Returning to the past pattern of multiple, scattered townships which will be black *de facto* rather than *de jure* would help integrate Kimberley's populace, would check the latent problems of massed urban living, and would open a window for enhanced spatial differentiation according to socio-economic criteria. Presently, the temptation is to build on land which the City Council has already bought contiguous to group areas, and to service it inexpensively by extending existing reticulation networks. It would be desirable to crack this racial mould as soon as possible and to acquire land for residential development beyond current mining and municipal limits to the east of the city.

Note

1. Archival material has not been cited; details may be obtained from the author. Principal sources include the Kimberley Town Clerk's records and the Town Council Minutes (Cape Archives, Cape Town); papers in the Africana Library, Kimberley; the *Diamond Fields Advertiser*; documents in the Historical Papers collection, University of the Witwatersrand; four unpublished theses written in Afrikaans.

9

JOHANNESBURG

S. M. Parnell and G. H. Pirie

Johannesburg, 100 years old in 1986, is in many respects the primate city in South Africa. It is not the capital city nor the mother city, but it is the most important economic centre and is the heart of the most populous and prosperous urban agglomeration in the subcontinent, the Witwatersrand. This complex of 10 towns is home to some 20 per cent of the country's 34 million people, and stretches for 80 kilometres along what was once the world's richest gold-bearing reef, from Springs in the east to Randfontein in the west (Figure 9.1). Not surprisingly, the racial complexion and complexities of settlement in these mining towns followed a similar trajectory. This began with the compounding of migrant labourers by mining companies before the turn of the century, and developed into more elaborate schemes of residential segregation and apartheid that were sponsored by the central and municipal governments.

The detail of settlement patterns and processes in individual Witwatersrand towns differed in response to local conditions. Least important of the controls on urban growth was the physical landscape: on otherwise featureless grassland, modest natural ridges and then imposing golden dumps of mine-waste were the only topographic features to play a part in segregating the tiers and races of society. More important agents of variation were differences in the rate of urban growth, divergence in the economic base and the occupational composition of work-forces, and discrepancies in the political complexions of local governments in their relations with the central government. These socio-economic and political cleavages persist to this day, with the result that although the Witwatersrand is a belt of contiguous metropolitan activity, it is not a monolithic entity. Certainly, the processes that shaped the individual towns which were more discrete in the past are not the same. In this whirl of disparate urban histories, that of Johannesburg involved the most people and is the best researched; in the skein of interdependent urban futures, that of Johannesburg will be decisive.

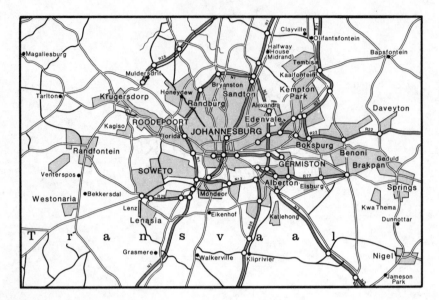

Figure 9.1   The Witwatersrand

Urban segregation

The residential geography of Johannesburg has never been free of racial considerations. The very earliest allocation of land made provision for separate
locations for Africans. More importantly, blacks were prevented by the Gold
Law of 1885 from owning and residing on land that had been proclaimed for
mining. The land shortage, and the ensuing overcrowding of black suburbs is a
recurring theme in the history of Johannesburg. So too is the periodic resettlement of black people. The first removal took place in 1904 when a severe
outbreak of plague was used to justify the destruction of the inner-city 'Coolie
Location' and the removal of the black population to a peripheral site at
Klipspruit, the nucleus of present day Soweto. Simultaneously the 'Kaffir
Location' was disbanded to make way for a cemetery for whites (Kagan, 1978).

The emergence of a racially pure lattice of residential townships in early
Johannesburg was undermined by the labour requirements of the mines, the
municipal authority, and white householders: the confinement of black people
to remote locations meant long, expensive and exhausting journeys to work
(Pirie, 1986) and was not conducive to reliable, regular and prolonged productivity. The mines soon discovered that round-the-clock transportation to the
mine shafts was eased by accommodating their all-male workforce at dormitories
on mine property. Similarly, the municipality found that if it were to provide
the city with essential services such as sewerage and refuse removal, its labour
force required easier access to the city than was possible from the municipal
location at Klipspruit (Pirie and da Silva, 1986; Pirie, 1988). Accordingly,
permission was granted for the accommodation of bona fide African employees

in compounds dotted about the town, and in servants quarters in the backyards of white residential properties (van Onselen, 1982).

Because the journey from Klipspruit was so long, many African people chose instead to live illegally in 'white' Johannesburg. After the First World War the number of Africans in Johannesburg rose dramatically as the economy diversified away from gold and as the families of migrant workers urbanized. Three forms of accommodation were available to those who scorned or did not qualify for servants quarters, compounds or housing in the municipal locations. First, the freehold areas of Sophiatown and Alexandra (Figure 9.2) offered wealthier blacks the privilege of purchase. The majority, however, were forced into rented rooms in these freehold ghettos. A second, more attractive option in the 1920s was the Malay Location (Figure 9.2). Africans favoured the area because it was close to town. Even though they were not permitted to sub-let from the coloured and Indian standholders, many ignored the restrictions. As a third option, blacks rented accommodation from white landlords on commercial and industrial property in the centre of town (Figure 9.2). This too was officially forbidden, but a way round the restriction was to pretend one was employed on site (Parnell, 1988a).

Restrictions on African settlement in Johannesburg were intensified after 1924 when the Natives (Urban Areas) Act of 1923 was made applicable to the town. The second wave of segregationist removals had begun. Of the 150,000 Africans in Johannesburg in 1931, an estimated 50,000 were directly affected by the Johannesburg municipality's attempts to bar Africans from illegitimate occupation of white premises and from sub-letting rooms in the Malay Location (Trump, 1979; Koch, 1983). From then on the only legal addresses for Africans were the municipal locations, including the new Western Native Township where a few hundred houses had been built (Figure 9.2), and the freehold areas of Sophiatown and Alexandra which were not included in the 1923 restrictions. As the available housing stock nowhere near satisfied the demand for shelter, blacks defied the law and rented space in the racially and culturally diverse slums of the central city. For once, the financial interests of the slumlords and the preferences of the black working-class coincided: in court the rentiers argued successfully that the Johannesburg City Council was not entitled to remove African tenants until it had provided alternative accommodation. Enforced segregation was thwarted temporarily, but racially integrated slums proliferated.

The shortage of working-class housing in Johannesburg grew more acute during the depression years of the late 1920s and 1930s. Racial mixing in cheaper neighbourhoods increased when the number of poor whites migrating to the city escalated. Urban living conditions deteriorated as the expanding central business district encroached on residential properties. Johannesburg's golden jubilee of 1936 coincided with the implementation of a sustained anti-slum drive that was the guise for the third major initiative to enforce residential segregation. The hope of the 1930s was that slum clearances and public housing provision would fix racial separation indelibly (Parnell, 1989a).

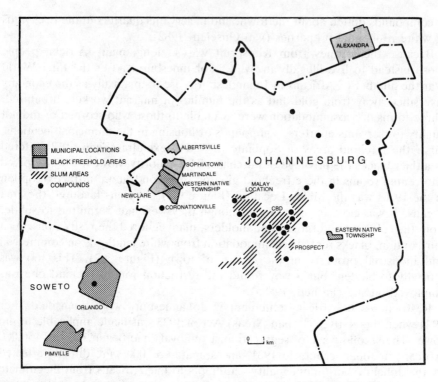

Figure 9.2   Black residential areas of Johannesburg prior to World War Two

The construction of public housing in Johannesburg between the two World Wars was ostensibly for all race groups, but in practice it affected mostly poor whites and Africans. Some housing for coloureds was built at Coronationville; Johannesburg's Indian community received no state housing assistance. The extensive public housing initiative of the 1930s was encouraged by the improved economic climate that followed South Africa's departure from the gold standard. The skeletal structure of modern, segregated Johannesburg dates from this inter-war period of active municipal intervention in the allocation, construction and management of houses for the poor.

In 1931 half Johannesburg's 400,000 population was white, of whom 8,000 were unemployed and 1,200 families were registered inhabitants of declared slum properties. The presence of thousands of poor whites in the slums of Johannesburg galvanized the slum clearances and public housing initiatives of the inter-war years. Although many loopholes in the Natives (Urban Areas) Act had been closed and the Johannesburg Municipality was systematically applying an amended form of the 1923 legislation, it was feared that the rate of African clearances would not stop racial mixing. Furthermore, as the Urban Areas Act only applied to Africans, coloureds and Indians were unaffected by its restrictions. In an effort to segregate and to rehabilitate destitute whites in Johannesburg, hostels and council houses were built to provide a subsidized and supervised living environment (Parnell, 1988a).

Although the attention given to the poor white housing problem was disproportionately great, it was not the sole concern of the city authorities (Parnell,
1988b). In 1932, houses were also built for Africans at Orlando (Figure 9.2).
The buildings were of the same design as those in white and coloured schemes
although the standards of finish were inferior. This inequality was less of a
concern than the distance between Orlando and the central city on the one
hand, and the rigid control associated with municipal tenancy on the other. In
consequence there were vacant houses at Orlando until the Second World
War. Africans who were evicted from their inner-city quarters by the Natives
(Urban Areas) Act or the Slums Act of 1934 preferred to shelter in the
freehold enclaves of Sophiatown and Alexandra or moved to the Malay Location
and Prospect Township where their presence, although illegal, was condoned
(Figure 9.2). So great was the influx to these overcrowded suburbs that in 1937
the Johannesburg City Council decided to relocate the entire African population
to a new dormitory township south of Orlando. The anti-slum provisions of
1934 and the sanitary regulations of the freehold and leasehold townships were
invoked to accomplish the clearances (Hellman, 1949; Morris, 1981a).

In Johannesburg, the segregation plans of the 1930s displaced the racially
integrated slums but failed to remove them. The local authority provided new
housing for white slum dwellers but failed to do the same for blacks. As a
result, places such as Sophiatown, Prospect Township and the Malay Location
became dumping grounds for black people evicted from white Johannesburg.
In the Malay Location the pressure on housing caused considerable resentment
and hardship for working-class coloured and Indian people. Coloureds in
particular complained that it was no longer possible to obtain affordable accommodation in Johannesburg. Shortages were aggravated by the demolitions
preparatory to the construction of commercial sky-scrapers and buildings for
burgeoning light industry. In an effort to secure some privileged access to
housing, the coloured community itself called for the establishment of segregated
residential areas. The Coronationville public housing scheme (Figure 9.2) was
not yet ready for occupation, however, and so a freehold township for coloureds
was proclaimed on the north-west edge of the Johannesburg municipal boundary.
As had happened at Orlando, the remoteness of this new township, Albertsville
(Figure 9.2), discouraged initial settlement. Furthermore, the development did
little to arrest squalid housing: streets in Albertsville were untarred and no
electricity or sewerage was laid on (Abrahams, 1963).

The Johannesburg housing crisis was nearing critical proportions even before
1939, but it erupted in full force when the population of the city doubled
during the Second World War. Houses in municipal locations were filled
beyond capacity, with lodgers in many of the units. In the inner-city, as well as
in the Malay Location, Sophiatown and Alexandra, houses were overflowing.
Rental for the most primitive shelter in these places was beyond the meagre
wages of the black working class. Thousands of people, particularly women
who did not qualify for access to municipal shelter, spilled out of the slums and
the backrooms of the municipal locations to squat on land south-west of
Johannesburg that the local authority had purchased for an African dormitory

town (Stadler, 1979; French, 1982; Bonner, 1989). These well-organized and strongly led squatter groups threatened the stability of the established African townships and obliged the municipality to give priority to rehousing squatters in the post-war period (Shorten, 1963; Lewis, 1966; Mandy, 1984; van Tonder, 1990).

Housing provision for whites in post-war Johannesburg was directed mainly at returning soldiers and was responsible for the growth of several fairly high-density suburbs within reasonable proximity of the central city. Forty years later, these small homes are the core of the gentrification movement. Unlike the pre-war period, the stress was on home-ownership and not on rent subsidization. Notwithstanding the shortage of building materials the white housing backlog was largely overcome in the first decade after the War.

The shortfall in black housing was more severe: in 1946 approximately five times more houses were required for blacks than were needed for whites. The number of sub-economic houses required for coloureds was 4,000 and for Indians was 3,000. Africans needed 42,000 units (Hellman, 1949). Despite these massive requirements, Africans were consigned less than 2 per cent of all building materials in 1946 and construction began slowly. Building of the standardized 'matchbox' houses was delayed further by a shortage of skilled construction workers (Crankshaw, 1990). In due course, the Johannesburg municipality sanctioned the construction of Dube and the laying out of Moroko, and approved extensions to Orlando and Jabavu. In doing so, racial segregation in Johannesburg was taken to an advanced stage even before the 1948 National Party victory heralded the introduction of apartheid in Johannesburg.

Urban apartheid

The target of racial residential separation in Johannesburg was the same in the apartheid era as it had been before, although the procedures of division were made more formal and were applied more vigorously. As in other cities, the notorious Group Areas Act was pressed into use, and in Johannesburg, was supplemented by yet other instruments. Events in the Western Areas are a good example. The removal of the slum areas to the west of the city had been proposed by the local government as early as 1937, but it was not until the 1950s that the bulldozers and trucks began their work amidst vocal but ineffective resistance. In moving 58,000 Africans to Meadowlands and Diepkloof (Figure 9.3), the government rode roughshod over the local authority's recommendation that the Western Areas be eliminated slowly as a slum clearance programme. In 1951, in a flagrant violation of local autonomy, the government circumvented the unco-operative stance of the City Council by enacting the Natives Resettlement Act which authorized removals despite the wishes of the local authority. The African sub-tenants of the Malay Location, by then renamed Pageview, were relocated simultaneously. On the wastes of Sophiatown there arose a low-income white suburb whose name celebrated the triumph of apartheid (Lodge, 1981; Hart and Pirie, 1984; Pirie and Hart, 1985).

In addition to effecting a ruthless programme of removals, the ideology of

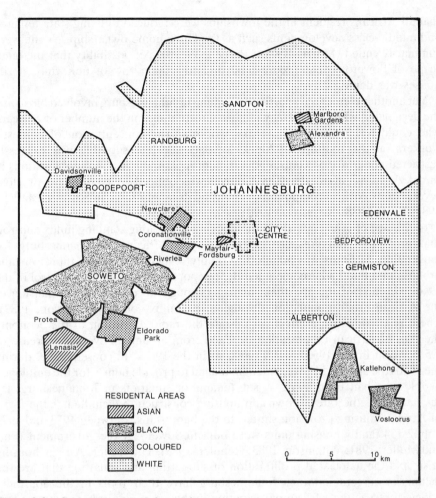

Figure 9.3   Group Areas in Johannesburg

apartheid was imposed inside the African townships. In the South Western Native Townships, now known officially by the acronym 'Soweto', the customary processes of public housing allocation were adorned by gratuitous reference to a person's supposed tribal group. In accordance with the belief that Africans were only temporary urban residents, Sotho and Nguni people were allocated homes in different areas so as to preserve rural links and enhance the schooling of school children in their mother tongues (Mashile and Pirie, 1977; Pirie, 1984a, 1984b).

Governmental interference in township affairs in the 1950s took yet other forms. An integral part of apartheid planning involved undermining African tenure. In accordance with the notion of the African's temporary status in white urban areas, title to township property was reduced to 30 years in the early 1950s, and was revoked completely in 1968. The immediate result was that all Africans resident in the townships were reduced to being state tenants, and middle-class developments such as the Dube home-ownership scheme were summarily ended (Morris, 1981a; Parnell, 1989b). All possibility that the landscape of Soweto would progress beyond row upon row of uniformly sterile houses was denied.

Yet another element of apartheid planning in Johannesburg involved bleaching the designated white suburbs by imposing restrictions on the number of Africans who could reside there as essential workers. The concentration of domestic workers on the rooftops of blocks of flats in white suburbia was expressly restricted under a 1954 central government measure which limited 'locations in the sky'. In addition, the City Council was obliged to consolidate or close several of its African compounds, rehousing its labour force in Soweto (Mather, 1987; Pirie, 1988; Rogerson, 1990).

In this age of grand apartheid the policing of the long standing influx controls that prohibited the presence of unemployed Africans in Johannesburg for longer than 24 hours was adhered to rigidly. Frightened by the rapid wartime urbanization that had accompanied the temporary lifting of influx control in the 1940s, the first of the apartheid managers sought to contain the expansion of the African urban population through strict influx controls (Hindson, 1987). These restrictions were effective: despite the rapid expansion of the economy the rate of African urbanization declined from an annual average increase of 4.5 per cent in the 1950s to 3.9 per cent in the 1960s. In consequence, during the 1950s the apartheid government managed to provide homes for the employed people who had been living as sub-tenants or squatters in Johannesburg. By the late 1960s the rate of township housing construction dwindled as the focus of African housing provision shifted to the bantustans. Between 1973 and 1979 only 2,734 family housing units were built in Soweto by the government (Fair and Muller, 1981; Wilkinson, 1983; Hendler, 1988). The chronic African housing crisis and the associated proliferation of sub-letting and squatting that are the hallmarks of contemporary Johannesburg have their roots in the apartheid policy of restricting African participation in the expanding Johannesburg economy and preventing large-scale building of new housing.

Africans were not the only group affected by apartheid surgery in Johannesburg, nor were they the only group whose remoteness from places of work (and therefore whose potential absenteeism and boycott action) (c.f. Stadler, 1981; Pirie, 1983) was eased and dampened by employer-based transport subsidies; slowly coloureds and Indians were also drawn into the scheme. Among the coloureds affected were people who, newly settled in Albertsville, overnight found themselves living on land proclaimed for whites. Unlike Africans who were denied urban tenure under apartheid, coloureds who could afford to were given the opportunity to purchase homes in the middle-class district of Bosmont. Less affluent families moved to the high-density, low-quality council flats at Riverlea (Figure 9.3) or occupied the rather derelict vacant units in the old Western Native Location, by then renamed Westbury to denote its new racial coding. Living standards in some portions of the coloured townships are amongst the most desperate in all Johannesburg (Brindley, 1976).

The Group Areas Act was particularly harsh on Johannesburg's Indians, the majority of whom lived in Pageview (the old Malay Location) and in the neighbouring areas to the immediate west of the city in Burgersdorp and Fordsburg. Proposals to relieve the pressure in these overcrowded areas were made soon after the Second World War and the City Council actually purchased land at Diepkloof for an Indian township. Nothing ever came of the plans, however, and relocation only became a real possibility when the likely designation of Pageview as a white residential zone was announced in 1952. At first the City Council seemed to assent to the implications of a forced relocation of Indians from the inner-city, but it recanted later and opposed the destruction of Pageview. Shortly, however, the Council itself publicized the establishment of Lenasia, a new Indian area beyond Soweto (Figure 9.3).

The foundations of Lenasia were laid in 1963, 10 years after the relocation of Pageview Indians was first rumoured. The removal of residents from Johannesburg's oldest black ghetto to new homes 30 kilometres from central Johannesburg was undertaken in stages (Randall, 1973). The residents from a decayed section of the old Malay Location were rehoused first. By 1968 20,000 people lived in Lenasia. Later, in the early 1970s once many Indian businesses had been installed in shops in a new shopping plaza erected by the City Council, the traders who resided in the northern section of Pageview were relocated. Many petitions for Pageview to be declared an Indian group area were ignored by the government. Only a court interdict in 1986 stayed the eviction of the final 100 families on the grounds that alternative accommodation was not available. Despite resistance to the principles and practice of segregation, Lenasia flourished because of the acute housing shortage of the previous decades.

Shuffling the residents of Johannesburg according to their race ended integration and achieved racial residential purity in the first phase of urban apartheid. Once the initial quilting had been sewn, it was easy to maintain by ensuring adherence to racial conventions in the proclamation and occupation of new townships. Ever since the 1960s, racially segregated housing developments, whether public or private, ensured the entrenchment of racially exclusive residential areas in Johannesburg. Those who dared contravene group area restrictions, faced imprisonment or a fine. In the black townships of Johannesburg, apartheid perpetuated itself and was taken to its ultimate conclusion when the administration of Soweto was taken out of the hands of the Johannesburg City Council's Native Affairs Department and given over to the ill-conceived and under-financed Administration Boards in 1973 (Frankel, 1979; Grinkler, 1986).

Once administrative control of Soweto was removed from the Johannesburg City Council the fiscal base of the township was undermined. Traditionally, Johannesburg subsidized the Soweto Native Revenue Account, and without this support the domestic rent and service charges were insufficient to finance the maintenance and upgrading of Soweto (Bekker and Humphries, 1985). Increased rentals levied by the West Rand Administration Board and its successor the Soweto Council met with hostility and active opposition. An effective rent boycott lasted for five years and was only resolved in 1990 with the State writing off hundreds of millions of Rand in rent arrears. Calls for the reintegration of Soweto and Johannesburg into a single rate base so that Africans would also benefit from the higher revenue derived from commercial and industrial rates and taxes are receiving a more sympathetic hearing from the new Democratic Party controlled City Council.

As the local government gears itself for desegregation and a more equitable distribution of Johannesburg's considerable resources it is confronted with the awful legacy of apartheid. Perhaps the most damaging inheritance is the shortage of land and housing in the black townships. Soweto's landscape bears the imprint of the apartheid government's first endeavours to segregate African residence, to enforce ethnic changes, suppress the emergence of a middle-class and to restrict the number of Africans living in Johannesburg. Alexandra, Johannesburg's second African dormitory town, is equally, though differently, scarred by apartheid management. In 1963 the Nationalist government recanted its earlier decision to demolish the freehold location, and instead declared it a 'hostel city' for migrant workers. The status of 'Alex' has recently been amended in line with the new reformist apartheid dispensation (Jochelson, 1990).

Reforming apartheid

During the 1970s Johannesburg residential development was more or less a passive echo of the ascendant ideology of racial separation. However, in time residential patterns in the city came to reflect the decline of apartheid and even began to hasten it. Although the process of desegregation has been most rapid in Johannesburg, the changing character of the city is also a response to national political shifts. Apartheid reforms followed the 1979 Riekert Commission's pronouncement that Africans were crucial contributors to the white urban economy. Riekert recommended changes in influx control and the ending of job reservation, and suggested that the state should foster rather than repress the African middle class. These policies have had far reaching consequences in shaping the Johannesburg landscape. In line with the recognition of African permanence in South African cities, home-ownership is now encouraged. The face of Soweto began to transform slowly after the government announced in 1983 that it would dispose of its houses in a gigantic sale. More than 100,000 units were available for purchase in the greater Johannesburg region. Most of the houses were in African townships, and lesser numbers in coloured, Indian and white areas. Occupants were given first option to buy, but it was hoped that a fully fledged private housing market would soon replace the ubiquitous state rental sector.

In the African and coloured townships house sales were sluggish at first. Despite prices being greatly discounted, they were out of reach of the vast majority of tenants. Those who might have been able to bear the expense were put off by the terms which, for Africans, featured 99-year leasehold and were not based upon freehold occupancy. Some active resistance also developed around the argument that it was absurd for people to buy homes whose initial cost had already been recovered through years of rental payments. In the first five years of the great sale, only 15 per cent of houses were purchased and 47,000 houses in Soweto remained state property (Mashabela, 1988). Progress toward the development of a middle-class housing market was nevertheless quicker in Soweto than in many other South African townships. The de-racializing of many white collar and professional occupations in commercially progressive Johannesburg gave Sowetans a financial headstart: by the end of the 1980s home ownership rates were more than twice the national average of 15 per cent (Hendler, 1989). Taken together with houses constructed by the private sector, approximately 40 per cent of the formal stock in Soweto is now individually owned.

Although some of the emergent African bourgeoisie live in the original township houses, extending and renovating them when they can, an increasing number of Sowetans opt to purchase newly constructed homes in one of the élite areas of Soweto, one of which boasts the name 'Beverly Hills' (Mather and Parnell, 1990), or in smart new sections of Alexandra township. The financing of construction in African areas was eased in 1986 when building societies were allowed to bond properties even though residents only possessed a 99-year lease. Finally, in 1987 freehold for Africans was sanctioned. The impediments to the development of a fully fledged housing market inside African townships had ended, even though the Johannesburg housing market as a whole officially remained barred to blacks.

In practice, despite group area restrictions, black people had been penetrating 'white' Johannesburg from the mid-1970s, slowly transforming it into what, by 1990, was probably the most racially integrated city in South Africa. Least noticeable to start with were the handful of black executives who began to infiltrate certain of Johannesburg's élite white suburbs. To the dark-skinned foreign dignitaries and prominent black churchmen who had a toehold in 'white' suburbs were now added the high-ranking black employees of the large corporations. Some of these men and women qualified for company houses, perhaps during attendance at courses, or while on transfer, or so that they could entertain. Other black executives began renting their own homes and purchasing property, working either through a white nominee or by forming a company that masked the racial classification of the owner.

By its very nature, the extent of the illegal black penetration of low density, élite 'white' suburbs has been slight. By contrast, the trickle of blacks into the high density, inner city districts of Johannesburg in the mid-1970s turned rapidly into a flood. At the beginning of this 'greying' process when the borders between black and white were first blurring, the old blocks of flats in the central city and the newer high-rise blocks in adjacent Hillbrow were home to a mixture of blacks and whites. After a phase of racially-motivated evictions and Group Areas Act prosecutions (Elder, 1990), the sheer weight of numbers began to tell, and soon apartheid had been stood on its head. By the end of the 1980s blacks had claimed the central city. Hillbrow and Joubert Park, once home to a shifting population of young whites as well as a sedentary set of poor whites of all ages, had become an over-crowded, crime-ridden black ghetto (van Niekerk, 1990a, 1990b). The process of physical degeneration was aggravated by the illegality of the black incursion. Landlords were uncertain about the future of their investments and let building maintenance lapse. Tenants eager for any kind of shelter paid inflated rentals and had no legal status to protest about contraventions of rent agreements.

The initial movement of black people into central Johannesburg was primarily a response to the grave shortage of accommodation in the black townships (Pickard-Cambridge, 1988a). Rentals in the 'white' city were higher, but could be offset against the lower costs of the shorter journey to work, and compensated by the accessibility of a wide range of shopping opportunities. Under the 1923 Natives (Urban Areas) Act, retailing and commerce in Soweto was stunted so that in the late 1980s there were only 1,312 registered businesses to serve a population in excess of 1.5 million (Mashabela, 1988). The network of hawker stalls and small backyard stores that sprung up only filled the gap with essential consumer items. Other reasons for moving into Johannesburg were to escape the violence in the townships and to ease travel to and from workplaces such as offices which were steadily decentralizing to peripheral sites in Rosebank, Sandton, Edenvale and Randburg (Figure 9.4). The attractions of a centrally located home increased as the liberalization of South Africa's racial laws gradually created a more hospitable environment in which public amenities such as libraries, cinemas, parks, restaurants, hospitals and public transport were desegregated. Similarly, in the face of a hiatus in state-sponsored black education, a central location offered proximity to desegregated private schools in 'white' areas (Parnell and Webber, 1990).

On the flanks of central Johannesburg, to east and west respectively, coloured and Indian people were the dominant agents of residential integration in the 1980s. In the east, the working-class suburbs of Doornfontein and Bertrams had long been a zone in transition. There is some evidence that coloureds may be returning to sites from which they or their parents were evicted during slum clearances in the late 1930s (Pickard-Cambridge, 1988a; see also Rule, 1988, 1989). To the west, in Mayfair especially, the migration of Indians into white working-class suburbs also contains elements of a return to roots which the removals from Pageview severed but failed to kill (Randall, 1973; Fick, de Coning and Olivier, 1988). Many merchants retained their commercial interests in the area when first their homes were displaced to distant Lenasia, and the sheer convenience of Mayfair is a great attraction. Unlike the varied tones of an earlier Indian presence, however, contemporary Indian gentrification of the western city has given it a uniformly upper middle-class character and has fixed correspondingly high land prices. Wealth continues to be the decisive filter in the gravitation of Indians to other parts of Johannesburg, such as Marlborough, a recently established prestige township abutting Alexandra, where the basic elements of the apartheid city are preserved at the edges of the opulent 'white' northern suburbs (Figure 9.4).

Changes in the geography of black residence in Johannesburg merged with political and economic developments in the 1980s to provoke fluidity in 'white' suburbs and also to generate conspicuous changes in the suburban built environment. The white housing market is acutely sensitive to fluctuations in South

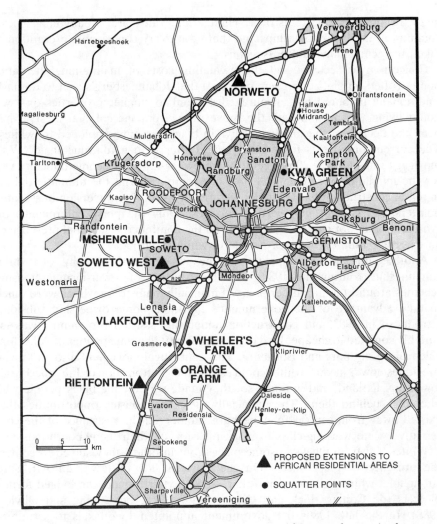

Figure 9.4   Squatter areas and proposed extensions to African settlement in the greater Johannesburg region, 1990

Africa's volatile political climate (Hart, 1990). In the last five years of social and political flux, the character of white middle-class suburbs has shifted in response to the privatization of security. Theoretically the South African police force is responsible for controlling suburban crime. However, police personnel have been increasingly deployed in township events and are noticeably absent from suburban streets. In the same period rising unemployment and the abolition of influx control have fed white fears over their security. Commercial companies now offer a 24 hour 'armed response' service to protect domestic premises against theft and unlawful entry. In the affluent white northern suburbs the logos of these security firms as well as the boards of 'neighbourhood watch' societies are prominently displayed on the high perimeter walls that characterize many of the wealthiest districts. High crime rates are, if anything, more common in the black townships and there too every conceivable precaution is taken to secure individual possessions.

Concern over security and the escalating costs of maintaining the large homes (often on plots of an acre or more) that Johannesburg's well-to-do have traditionally favoured has encouraged a trend to smaller properties or town house living. Associated with the move away from the colonial lifestyle has been the decline in the number of live-in black servants employed by whites. The current practice is for a daily char rather than a full-time maid. With mortgage rates at over 20 per cent, home ownership has become difficult even for the middle-classes. Financial pressure combined with the dramatic fall-off in rental construction has created a lucrative trade in the conversion of servants quarters and garages into garden cottages. Only the persistent reliance on African women for child care has ensured the continued residential presence of an extensive number of black people in white suburbia.

Another marked development in the residential geography of 'white' Johannesburg during the 1980s was the eruption of lower-income housing schemes around the urban periphery. Financially hamstrung whites were quick to take advantage of the government's first-time home ownership subsidies introduced in 1983, but construction under this assisted programme was invariably confined to cheaper land in the neighbouring municipalities of Alberton, Edenvale, Randburg and Boksburg. This centrifugal movement also had racial overtones: lower income whites could escape the ghettoization of Johannesburg's low rental flatland, and in doing so, they made way for the black people who filtered in behind them. The financially attractive housing opportunities that existed away from central Johannesburg submerged the resistance of the conservative white working-class to black invasion of 'their' neighbourhoods.

Whites fled from blacks in some places, but in others they stayed to contest the presence of black faces. On the western side of town, an Afrikaans stronghold, white vigilante groups harassed Mayfair Indians and lodged formal protests (de Coning, Fick and Olivier, 1987; Fick, de Coning and Olivier, 1988). In the late 1980s the government appointed 13 officials to scrutinize group area infringements in Johannesburg, and they handled almost 200 complaints. No prosecutions were made. In the new era of political reform and negotiation, prosecutions under the Group Areas Act became discretionary rather than mandatory, and there was even a stay of convictions for previous group area contraventions.

The first-time home owner subsidies from which whites benefited were not race-specific and also applied to blacks who wished to buy their first house. Whereas approximately 80 per cent of whites in Johannesburg could afford a subsidized house, the proportion of Africans who could even afford to use the subsidy was only in the vicinity of 20 per cent. If the impact of the subsidies in African townships was limited, it was nevertheless obvious. The speculative housing projects of commercial contractors who cater to the subsidized home-owner market contrast sharply with the housing estates erected by the government in the 1950s. In places, Soweto and Alexandra are losing their bleak uniformity: homes have varying architecture and setting; there are houses with tiled roofs and garages. These changes to the face of the African townships are likely to persist and deepen. Because the State has effectively abdicated its traditional role as housing provider, private sector initiatives are now the only real means of expanding the existing formal housing stock of Soweto. The problem with the first-time home-ownership scheme in African areas is that only an estimated 18 per cent of the population could afford a formal house (only 8 per cent could afford the same house without State assistance). In an effort to lower the entry point into the African housing market the Urban Foundation has recently launched a massive fund to lend money for units costing half the amount that can be secured via the government's first-time home-building scheme.

Since the allocation of 'own affairs' responsibilities under the tricameral parliament of 1984, there has been a marked increase in State funded housing within coloured and Indian group areas. Despite paying lip service to the importance of privatizing housing, sections of Johannesburg's coloured and Indian population now rely more on the State for shelter than they have ever done (Hendler, 1989). Although there is a continual shortage of affordable shelter, especially in coloured areas, widespread squatting remains an exclusively African phenomenon.

The development of middle-class suburbs in Soweto and Alexandra in the 1980s occurred amidst a worsening shortage of low-income housing. Shanties sprang up to accommodate the homeless in backyards throughout the townships and, for a time, were relentlessly demolished by the authorities. In the long run the tide of events was once again irresistible. Estimates are that one million people are currently sub-tenants in Soweto, an increase of 22 per cent in two years. Aerial photographs reveal a minimum of 640,000 backyard shacks in the formal townships of the Witwatersrand. Even the tightly packed warrens of corrugated huts failed to give everyone shelter, however, and in the late 1980s the phenomenon of squatting on open land re-appeared. Now, in 1990, some 2.5 million people are thought to occupy informal housing around Johannesburg, 50,000 of them in squatter shacks (Mashabele, 1990). If these estimates are correct half of the Africans living on the Rand are accommodated in informal rather than formal housing. At infamous Mshenguville in Soweto, and at Kwa Green on the Alexandra border, sub-tenants have been squeezed out of back-yards on to nearby vacant land. At places such as Vlakfontein and Wheeler's Farm, whites on smallholdings have let sites to squatters. In an effort to eradicate this new generation of 'black spots' the government itself established site-and-service camps which are dumping grounds for squatters evicted from other localities. One such consolidated squatter camp is at Orange Farm to the south of Johannesburg (Figure 9.4) (Mashabela, 1988; Crankshaw and Hart, 1990). Several similar schemes elsewhere on the Witwatersrand have been proposed.

The shortage of housing in Johannesburg's formal African townships is partly the result of failure to incorporate any new land into those townships. Officially there was no need to enlarge dormitory towns in the era of rigid influx control. Expansion has become imperative so as to relieve gross overcrowding. Unfortunately, sites are in short supply. Those that were originally selected for segregated townships because they were remote are no longer isolated: Alexandra is all-but surrounded by industrial and residential areas; Soweto is bounded by mine-owned land to the north and the east; Lenasia backs on to mine land to the south. Extension is limited, but not yet impossible (Figure 9.4). Plans are in hand to extend Soweto westwards, and the township is expected to double in physical size in the next decade. In addition, approval has been given to the proposal for a 'second Soweto' 50 kilometres south of Johannesburg at Rietfontein, despite severe misgivings about the dolomitic soils (Mashabela, 1988). This siting has drawn less criticism than did earlier proposals to duplicate Soweto on the peri-urban fringe of northern Sandton and Randburg. The so-called 'Norweto' project (Figure 9.4) drew howls of anguish from local whites. Plans for a mass African housing project were shelved, but in time they were replaced by plans to establish a Free Settlement Area which would comprise predominantly owner-built homes (Elder, 1990). One option for the proclamation of new residential areas may arise in future if title to mining land close to the city centre is negotiated as part of a new deal with big business in the emergent political order. Johannesburg is unusual among world cities in having vast swathes of centrally situated derelict land awaiting redevelopment: geological instability will probably confine construction to low-rise, low-density housing.

Advanced plans for new segregated townships, and the reality of the existing mosaic of residential apartheid, are likely to guide future residential growth in Johannesburg rather than constrain it. In post-apartheid South Africa the rigidity of group area restrictions and the tokenism of free settlement areas will disappear. Eventually, the geography of settlement will reflect market forces more closely. In this regard, Johannesburg may well set the national trend irrespective of whether or not the city assumes a more prominent role in national government. Already the City Council has indicated its intention to declare the entire municipality a free settlement area, thus pre-empting the government's promise to remove statutory residential segregation. Should the neighbouring liberal councils of Sandton, Randburg and Bedfordview follow suit, enforced residential segregation would effectively be undermined.

Although the process of desegregation has already been unleashed in Johannesburg, integration in a post-group areas, post-apartheid South Africa will undoubtedly be more common even if it is not the norm. In a vibrant urban milieu, heartland of the national and regional economy, the black middle-class in particular is likely to expand rapidly and to upgrade the physical fabric of the townships steadily. The integration of 'white' and 'black' Johannesburg into a single tax base under unified municipal government will certainly allow increased expenditure on 'township' services. Black residential relocation to suburbs which were previously reserved for whites may also be expected to intensify. The wealthiest in society will always seek opportunities to invest in high-grade property; others will wish to relocate for reasons of accessibility to work, schools and shops.

In relatively affluent Soweto, departing residents may have more opportunity to sell their township houses than would be the case in smaller and poorer locations where the housing market is more saturated and where the prospects of significant in-migration and house filtering are dim. Correspondingly, home-owners at the apex of the urban hierarchy probably have more opportunity to purchase homes than their counterparts in smaller centres: householders in Johannesburg's plush and leafy northern suburbs are relatively liberal in their political outlook and are unlikely to barricade their territory against blacks who can afford to join them and who will subscribe to the same values. Equally, politically conservative whites can easily relocate inside the metropolitan area without the kind of disruption to occupation, lifestyle and social relations that creates a defensive inertia in people whose only relocation option involves a long trek from a small town.

Developments in the formal housing market will impinge on informal settlements that have already established themselves as a permanent feature of the Johannesburg landscape. As the city expands and peri-urban and agricultural land is developed farm labourers, few of whom have access to the skilled urban labour market, will migrate to peripheral squatter camps. Families of miners, once barred from urban areas such as Johannesburg, have already begun to move to the reef and this trend may be expected to continue. The recent expansion of the number of squatters has prompted several illegal land invasions. Depending on the housing policy of the new political dispensation and the amount of land released for low-income housing in the immediate future, this may become more common. What is certain is that the easing of the rigid control of apartheid rule will bring to the cities much of the poverty that has hitherto been hidden in the bantustans.

In the new South Africa, housing in Johannesburg and environs will encompass the entire spectrum from obscenely ostentatious villas to grindingly spartan slums. These homes will always be apart, but segregation will be conducted along lines of social-economic class rather than skin pigmentation, and explicit racial segregation will have been outlawed.

10
PRETORIA
P. S. Hattingh and A. C. Horn

From its founding in 1855 to house the government of the dispersed Boer communities of the Transvaal, to its present function as the administrative capital of the Republic of South Africa, the growth and development of Pretoria has been closely linked to spatio-political ideology and decision-making. As an integral part of the Pretoria-Witwatersrand-Vereeniging (PWV) urban complex (the largest conurbation in Sub-Saharan Africa) it serves as the northern passage linking the urban system with the Bophuthatswana and KwaNdebele bantustans. In essence it shares the spatial structures of other South African cities which also display the superimposed characteristics of apartheid and grand apartheid, but due to some specific spatio-constitutional developments, it is arguably the most complex.

The aim of this chapter is to highlight the evolution of the Pretoria spatial and politico-administrative complex, its daily rhythm of ebb and flow along ethnic lines, and to present some post-apartheid scenarios of the city.

The evolution of Pretoria as an urban complex

Over thousands of years the region commonly known as the Pretoria area was formed into a succession of sedimentary ranges and well watered valleys inhabited by a variety of species including early man. Over the past few centuries, however, the area accommodated the sequential arrival of a number of civilizations such as those of the Ndebele, Sotho, Zulu and Europeans who, in their struggle for supremacy over this piece of sought-after land, transformed the human settlement pattern from a pastoral to an integrated pulsating complex of First and Third world urban forms, irregular spatio-ethnic patterns and diverse politico-administrative structures that today comprise the Pretoria functional region. Such is the complexity of this urban conglomerate that it does not fit into the framework of any relevant urban model such as 'colonial city'

(Lowder, 1986), 'segregated' and 'apartheid cities' (Davies, 1981; Western, 1986; Simon, 1989a), 'dispersed city' (Ginsberg, 1961; Burton, 1963) or the 'separate city system' (Olivier and Hattingh, 1985) and we therefore prefer the label 'multiple city system' for the prevailing state of Greater Pretoria at the end of the apartheid era in South Africa.

The White Afrikaner pioneers settled in the Apies River Valley (as the area was initially known) in about 1840 and gradually the amalgamated Ndebele and Sotho tribes who had been in the area for over three centuries (Junod, 1955, p. 65) had to give way. Pretoria was proclaimed a town in 1855 and soon (in 1860) became the capital of the 'Zuid-Afrikaansche Republiek'. The population increased from about 400 (mostly white) to at least 3,000 in 1877, when Pretoria was occupied on behalf of Britain.

By 1904 there were some 27,000 whites, 21,000 Africans and 4,000 Indians and coloureds in Pretoria. The African servants who found their way into the town had instructions to seek shelter, when night fell, at the nearby kraal of Maraba, a petty chief among the original Africans of the area. Marabastad, as it became known, developed almost unnoticed and soon displayed ghetto-like characteristics. The coloureds and Indians were also accommodated in Marabastad but segregated themselves from the Africans (Junod, 1955, p. 78). Despite worsening living conditions, the Marabastad 'location' and Schoolplaats adjunct were designated in 1905 for the exclusive occupation of 'natives' not living on the premises of their employers. A growing number of Africans also settled in dispersed peri-urban clusters such as Mooiplaats, Lady Selborne and in the Derdepoort area (Figure 10.1(a)).

By 1908, Pretoria consisted of a core of commercial and administrative functions as well as light industries and an oval ring of segregated residential areas determined by the succession of east−west stretching ranges of hills and valleys surrounding it. It seems that the ethnic-residential pattern of apartheid Pretoria had already by that time been established. Its becoming the administrative capital of the Union of South Africa in 1910 as well as the outbreak of the First World War (1914−18) led to the influx of a great number of people: in the 17 years from 1904 to 1921 the white population increased by 89 per cent to 51,000 whilst black numbers increased by 125 per cent to 55,000.

The party-political instability of the next four decades prior to 1948 was reflected in patterns of urban development. At first the attitude of *laissez-faire* prevailed but the state was soon confronted by the harsh realities in the black urban sectors brought to the fore by the enquiries of, amongst others, the Tuberculosis Committee (1914). The adoption of the Natives (Urban Areas) Act of 1923 introduced a stricter approach but really failed to achieve the intended objective of overall residential segregation.

Despite efforts to centralize relevant administration under the Department of Native Affairs, the control over the inflow of people into urban areas and their accommodation was very much the responsibility of the local municipalities. The Pretoria urban area in the 1930s and 1940s comprised four municipalities (excluding the dormitory town of Irene), namely Pretoria, Silverton, Hercules and Pretoria North and each adopted its own policies in respect of the housing

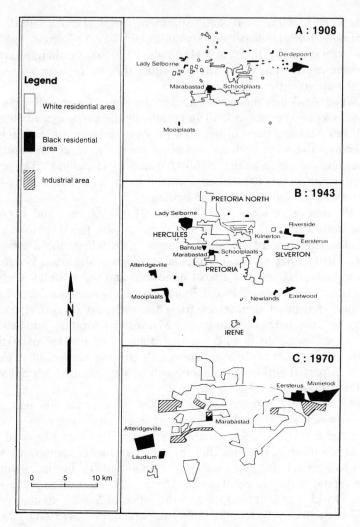

Figure 10.1   Residential distribution of ethnic groups in Pretoria: 1908, 1943, 1970

of urban Africans. The Pretoria Municipality (declared a city in 1931) officially accommodated blacks in the Marabastad-Schoolplaats Location as well as in Bantule, proclaimed in 1925 (Figure 10.1(b)). Atteridgeville was established in 1939 as another alternative. In contrast to this, the municipalities of Silverton and Pretoria North made no direct efforts to accommodate Africans within their boundaries since both these municipalities were located close to a number of peri-urban African clusters. The largest settlement of Africans was, however, to be found at Lady Selborne in the Hercules municipality. This was not a 'location' but a multi-racial freehold residential area that accommodated about 34,000 people including some 50 whites, 240 Indians and about 1,000 coloureds in the early 1950s.

Notwithstanding the measures taken, only 33 per cent of the estimated African population of 29,000 in 1930 was accommodated in the official locations with another 10 per cent residing on state property. The remainder resorted to squatting around the city. A census undertaken in 1952 revealed 16,939 squatters on farms west of Pretoria and 57,635 on farms east of Pretoria (Junod, 1955, p. 84). This figure of 74,574 rural squatters can be augmented with another 58,576 squatters in the urban area (excluding whites living in slum conditions), totalling 133,150 or 36 per cent of the population of 370,980 in the Pretoria Magisterial District.

Thus, broadly speaking, the 1943 residential pattern (Figure 10.1(b)) illustrates the consolidation of white suburbs around the urban core with the biggest concentration of blacks in the official locations, Marabastad-Schoolplaats, Bantule and Atteridgeville, the freehold area of Lady Selborne, squatter camps such as Mooiplaats, Kilnerton, Newlands, Eastwood, Eersterus and Riverside as well as in a large diversity of smaller clusters.

The 1970 configuration of ethnic residential patterns within the municipal boundaries of Pretoria (Figure 10.1(c)) very much illustrates the outcome of residential apartheid in Pretoria. African residential areas such as Mooiplaats, Schoolplaats, Bantule, Lady Selborne, Kilnerton, Eastwood, Newlands and Riverside had been demolished. Atteridgeville and Mamelodi (an African township constructed in the north-east) accommodated the bulk of the African population in Pretoria at that stage. Eersterus, a former African cluster adjacent to Mamelodi was redeveloped to accommodate the coloured population. An Indian suburb (Laudium) was built in the south-west with one of the hill ranges as a buffer between it and the African township of Atteridgeville. A small remainder of Indians, however, still resided in Marabastad. Kilnerton and Riverside, former African clusters, had been redeveloped as white residential areas and the same was done with Eastwood, Newlands and Lady Selborne after 1970. Schoolplaats currently houses the council's bus depot, and the Pretoria Technichon was recently partly built on the premises of the old Bantule location. None of the then numerous African clusters on the urban periphery still exist. The white residential area was consolidated over time and its coherence only disrupted by ranges of hills and streams and by industrial areas. The location of industries played an instrumental role in organizing the apartheid city. Three major industrial areas developed within the city limits. To this day two of these serve as buffer zones between the white areas and Atteridgeville/Laudium in the south-west and Eersterus/Mamelodi in the north-east. The siting of these industrial areas in relation to the black residential areas has obviously been deliberate.

To appreciate the current (1990) complexity of the multiple city system (Figure 10.2) some attention should be given to territorial separatism or grand apartheid as it is otherwise known. Although ethnic-territorial separatism is very much a historical legacy, the policy of territorial apartheid was formalized by the Nationalist government. Attempts to make this policy viable included efforts to consolidate the 'homelands' into units; encouragement of urbanization in the 'homelands'; and the creation of an economic foundation by the estab-

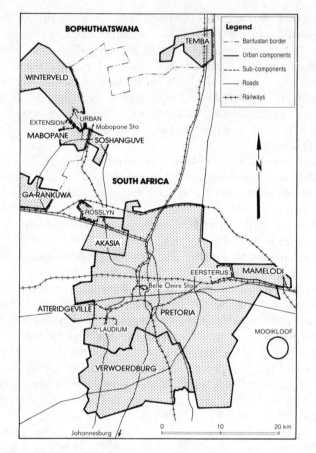

Figure 10.2   The multiple city of Pretoria in 1990

lishment of industries in and close to these African areas. In the Pretoria area
the process took off by the creation of Ga-Rankuwa (1962) and the government's
embarkation on a programme of clearing African peri-urban spots in the white
areas. Residential and territorial segregation led to the removal and resettlement
of about 155,000 people in the Pretoria area (Hattingh, 1975, p. 74).

  Since the late 1960s the City Council of Pretoria involved itself in the establishment of another urban centre (Mabopane) in the nearby bantustan and as a
result the creation of additional accommodation for Africans in Pretoria itself
ceased. The 'independence' of the Tswana nation state (Bophuthatswana)
which came into being in 1977 created additional problems further complicating
the matter. The Bophuthatswana authority was not prepared to accept the
Winterveld – a mostly non-Tswana squatter area – as part of the new state.
As a compromise Mabopane East, part of the newly founded Mabopane
township, was reincorporated into South Africa, renamed Soshanguve and
further developed by ethnic zoning according to strict apartheid policies to

accommodate the *So*tho, *Sha*ngaan, *Ng*uni and *Ve*nda peoples
Soshanguve reflects. This only slowed down the pace of url
compaction in the Winterveld and had even less impact on its et

Meanwhile the municipalities of Hercules, Silverton and Pretoria North were
also incorporated into Pretoria. The dormitory town of Irene in the south and a
new town annexure together became the city of Verwoerdburg. Residential
development for whites in the area between Pretoria and the Rosslyn border
industries gave rise to the establishment of Akasia as yet another municipal
component of the urban conglomerate (Figure 10.2).

The current composition of the Pretoria functional region may be summarized
by brief reference to four sectors. Firstly the Odi and Moretele Magisterial
Districts of the 'independent' bantustan Bophuthatswana accommodating the
formal towns of Temba, Mabopane and Ga-Rankuwa, each with an autonomous
city council, and the informal settlement of Klippan (the most southern portion
of the Winterveld area) which has official urban status. Furthermore, there is a
broad band of squatter settlements (including the rest of the Winterveld) from
Temba in the north stretching past Mabopane and Ga-Rankuwa to the western
border of the Moretele District. Secondly much of the self-governing bantustan
of KwaNdebele to the north-east of Pretoria also falls within the city's commuting
hinterland, and indeed its string of peri-urban settlements testifies to the
drawing power of (potential) urban employment. There has been considerable
migration into KwaNdebele since its designation in the 1970s as a 'homeland',
supposedly for the North Sotho people. As in the Winterveld, the migrants
include many members of the 'wrong' ethnic group, Third World would-be
urbanites who have moved as close to the cities as 'grand apartheid' allowed.

Thirdly the urban components of Pretoria itself include the three white
controlled urban council areas of Pretoria, Verwoerdburg and Akasia, the
latter incorporating the border industries of Rosslyn. There are also two autonomous African city council areas namely Mamelodi and Atteridgeville. The
coloured town (Eersterus) and the Indian town (Laudium) are administered by
Management Committees under the auspices of the relevant chamber of the
tricameral parliament but are spatially and structurally integrated within the
city of Pretoria. Fourthly, on yet another level, is the town of Soshanguve as an
annexure to the Bophuthatswana town of Mabopane but controlled by a
combination of South African state departments and provincial authorities.

As in all South African cities a significant number of blacks reside in rooms
on white residential premises. There have been numerous transgressions of
group areas regulations, but not sufficiently to produce any real 'grey areas'.
However, public amenities and business areas are progressively being opened
to all citizens. Although the 1988 Free Settlement Areas Act is probably
nothing more than a tactical manoeuvre within the constitutional tangles of
pre-1990 South Africa, several such possibilities in the Pretoria area (of which
Mooikloof is the most significant) are under investigation despite an orchestrated
fury from a section of the white establishment.

The daily rhythm of the Pretoria multiple city system

In this section the present functioning of the Pretoria multiple city system is discussed, highlighting three themes. Firstly, the ethnic composition of the population is dealt with in terms of its spatial distribution and socio-economic structure. Secondly movements interlinking home and daytime destinations receive attention. Thirdly, the ethnic spatial-time patterns over a 24-hour period prevalent in the system are described.

As elsewhere in South Africa uncertainty about the number of people in the Pretoria area prevails. Datakonsult (1989) estimates the urban population for 1990 as 1.3 million, a figure that will increase to 1.6 million in the year 2000. Another source (LHA, 1990) claims, as far as Africans are concerned, that the respective figures are 36 per cent and 53 per cent higher.

By using the estimate of Datakonsult (1989) as the basis for discussion, it appears that of the total of 1.3 million people, Africans comprise about 60 per cent and whites about 36 per cent of the population. Although two-thirds of the population are accommodated in the black components of the city, Pretoria city council (excluding Eersterus and Laudium) is the largest single local authority with just over half a million people or 39 per cent of the total population residing there (Table 10.1). Mamelodi (with an African local authority) was the second largest with close to 200,000 people.

The other components with white local authorities accommodate relatively small numbers of people: Verwoerdburg (61,000 people) ranks only seventh in terms of population size with Akasia (9,600 people) the smallest of the 11 components. Apart from Mamelodi (199,000) and Temba (55,000), the other African components of the city each have populations of close to or above 100,000 people. Laudium (Indian) and Eersterus (coloured) have populations of 20,000 and 23,000 respectively. Some 99 per cent of whites reside in the three white components of the city. Similarly 93 per cent of Indians live in Laudium with another 5 per cent in the white areas, especially Marabastad. Eighty-four per cent of the coloureds reside in Eersterus with another 11 per cent in the white areas and 2 per cent in both Indian Laudium and African Mamelodi. Eighty-eight per cent of Africans reside in African towns: 41 per cent in Mamelodi and Atteridgeville which are closer to the urban core, another 14 per cent in Soshanguve and the majority of the remainder in either the African towns of Bophuthatswana (33 per cent) or the white areas (11 per cent) and an insignificant 1 per cent in the Indian and coloured components. Thus, while the African, Indian and coloured components of the city are dominated by a specific ethnic group (for instance at least 99.8 per cent of the residents in the African areas are Africans), one in every five persons residing in the white municipalities is not white.

The economic spectrum derived from 1980 census tabulations (South Africa, 1980b) is represented by 10 urban components of which half are white sub-components or suburbs (Table 10.2). Comparatively speaking the annual per capita incomes (1980) were lowest in the African components varying insignificantly from R598 in Soshanguve to R635 in Mamelodi and R661 in Atteridgeville. By contrast to this there were meaningful differences in the income levels of the white suburbs with Danville representing a low annual per capita income of R2,260 and Waterkloof Park reaching R9,217. The annual per capita income of people in the first mentioned white suburb is thus approximately three times that of people in the African towns, nearly double that of the coloureds in Eersterus and also slightly more than that of Indians in Laudium.

In 1990 more than 45 per cent of African households in greater Pretoria had an income of less than R600 per month (LHA, 1990, p. 10). Also in the African sector 32 per cent of the 99,300 dwelling units were informal shack structures (Table 10.3).

The spatial complexity of the urban system with the socio-economic variety therein is indifferently kept intact by a comprehensive network into which the legacies of pre-apartheid, apartheid as well as grand apartheid transport systems are incorporated. Besides the influence of settlement policy, the road and rail links have also been determined by the geomorphological characteristics of the region. The transport network and the various systems functioning within it are distinctively utilized by the various ethnic groups. The whites and Indians, because of their higher economic positions and residential proximity to the urban core, mainly resort to private modes of transport and to a lesser extent make use of public bus services. Coloureds travel either by train, taxi or private vehicles while the Africans are mostly dependent on buses, taxis or trains.

Olivier and Booysen (1983, p. 126) estimated that there are more than 400,000 African commuters focusing on the Pretoria core. Of these almost 40 per cent live in the nearby towns of Mamelodi and Atteridgeville, another 40 per cent in the Winterveld/Mabopane/Soshanguve area, 12 per cent in Ga-Rankuwa, and five and one per cent in Temba and the KwaNdebele bantustan respectively (Morris, 1983, p. 18). The bulk of the commuter traffic is directed to the city centre and has influenced the functional structure of this core (Olivier and Booysen, 1983).

The Africans in the two towns closer to the core prefer the train and taxi as transport modes while bus services provide a supplementary mode for the settlements further afield. The public transportation of passengers has been tightly regulated (Clark *et al.*, 1988, p. 5) and currently there are three African bus companies operating in the area. The Mabopane-Belle Omre metro rail service is a major link between the city core and the urban components in Bophuthatswana and transports more than 40,000 commuters daily.

The containment of urbanization in South Africa has, in the case of Pretoria, led to extremely long travel distances, averaging 52 kilometres in one direction. The imposition of these long commuter distances on the poorest section of the community resulted in the subsidization of commuting and is yet another manifestation of the disparity between the layout, allocation and functional requirements of the South African city. Black commuters are therefore subsidized on the basis of compensation for unaffordable travel costs (ibid., p. 23). The total cost of bus subsidies in KwaNdebele actually exceeds the GDP of this impoverished bantustan (Lelyveld, 1986). Despite subsidization, commuters

Table 10.1   Estimated population of Greater Pretoria for 1990 and 2000

Component[1]	Year	Africans		Asians		Coloureds		Whites		Total
		Total	% of (x)	Total	% of (x)	Total	% of (x)	Total	% of (x)	(x)
Pretoria	1990	75,160	15.0					422,960	84.3	501,480
	2000	106,000	18.9					451,200	80.3	561,900
Mamelodi	1990	199,200	99.8							199,700
	2000	281,000	99.8							281,700
Atteridgeville	1990	115,170	99.8							115,450
	2000	162,500	99.8							162,900
Soshanguve	1990	107,590	99.8							107,840
	2000	151,800	89.8							152,100
Ga-Rankuwa	1990	100,110	99.9							100,210
	2000	141,200	99.9							141,300
Mabopane	1990	96,000	99.9							96,130
	2000	135,400	99.9							135,600
Verwoerdburg	1990	10,290	16.8					50,550	82.7	61,110
	2000	14,600	21.2					54,000	78.4	68,900

Temba	1990	54,850	99.9							54,930
	2000	77,400	99.9							77,500
Eersterus	1990					22,100	96.1			23,000
	2000					32,500	95.9			33,900
Laudium	1990			19,180	93.1					20,600
	2000			22,400	93.3					24,000
Akasia	1990	1,560	16.3					7,970	83.1	9,590
	2000	2,200	20.4					8,500	78.7	10,800
Greater Pretoria	1990	761,430	59.0	20,390	1.6	26,400	2.1	481,820	37.3	1,290,040
	2000	1,074,000	65.1	23,800	1.4	38,700	2.4	514,100	31.1	1,650,600

Notes
[1] Rank by population size.
[2] Only percentages above 9.9 included.
(*Source*: Datakonsult (1989).)

Table 10.2   Spectrum of selective incomes in Greater Pretoria (1980)

Urban component	*n*	Per Capita Income (per annum)
Soshanguve	62,476	R 598.00
Mamelodi	142,361	R 635.00
Atteridgeville	47,712	R 661.00
Eersterus	13,892	R 1,349.00
Laudium	14,726	R 1,951.00
WHITE SUBURBS		
Danville	8,607	R 2,260.00
Rietfontein	8,793	R 3,892.00
Menlopark	3,278	R 5,792.00
Waterkloof	3,097	R 7,867.00
Waterkloofpark	104	R 9,217.00

Table 10.3   African dwelling units in Greater Pretoria (1990)

Town	Formal	Shacks	Total	Hostel beds
Atteridgeville	10,200	3,800	14,000	11,511
Ga-Rankuwa	10,600	5,200	15,800	–
Mabopane	11,500	2,800	14,300	–
Mamelodi	17,000	16,000	33,000	10,932
Soshanguve	14,500	3,100	17,600	1,672
Others	3,800	800	4,600	3,908
Total	67,600	31,700	99,300	28,023

(*Source*: LHA, 1990, p. 6.)

spend more than five per cent of their income on transport. Furthermore, social and recreational trips are not subsidized and therefore the public transport system is socially ineffective.

By public transport a trip from Temba in the north to Verwoerdburg in the south takes about three hours. Because of the long travelling distances African commuters in Pretoria are severely disadvantaged. For 65 per cent of bus commuters, travel time exceeds 75 minutes (mean travel time is 106 minutes) according to Clark *et al.* (1988, p. 34). While commuting from Mamelodi and Atteridgeville to the city centre takes less than an hour, people from the outpost settlements in Bophuthatswana and KwaNdebele spend up to seven hours a day or even more on travelling and actually spend more time commuting on a daily basis than sleeping (*Pretoria News*, 30 May 1985). A sample from such an area shows that the average commuter leaves home at 04:00, travels for nearly four hours to work and only returns home at 19:50 (Fourie, 1985). The infrequency of public transport makes it impossible to plan the time of a journey effectively (van der Reis, 1983) and quite often results in people spending the night at work or at a transfer point such as Belle Omre station.

The introduction of the kombi-taxi as mode of 'peoples transport' offers a welcome alternative. It is more regular, reliable and comfortable than public transport and is more effective in linking the various urban components of this functional whole. It has, however, dramatically increased the volume of traffic and also poses inner city planning difficulties and wider management problems.

To comprehend the pulsation of the urban organism one must perceive its ongoing nature, through differentiated ebb and flow rhythms as well as the resultant cycles of human ecology. This is particularly true of South African cities upon which the different scales of apartheid are imposed, exaggerating the urban pulse to bizarre proportions. This can be illustrated by means of a time schedule of movements and associations in Pretoria over a 24-hour period:

03:00 Pretorians asleep.

03:40 First commuter train departs from Mabopane and arrives at Pretoria station on the southern fringe of the Pretoria CBD 53 minutes later.

04:05 The average worker in north-eastern Bophuthatswana leaves home for Pretoria.

04:09 The next train departs from Mabopane and arrives at Belle Omre Station on the northern fringe of the Pretoria CBD 47 minutes later.

04:15 The average worker in KwaNdebele leaves home for Pretoria, by bus. First commuter train departs from Ga-Rankuwa and arrives at Pretoria station 52 minutes later.

04:17 First commuter train departs from Atteridgeville and arrives at Pretoria station 22 minutes later.

04:24 First commuter train departs from Mamelodi and arrives at Pretoria station 31 minutes later.

05:10 The average worker in the border towns of Bophuthatswana leaves home for Pretoria.

06:00 The arrival of 13 commuter trains from Mamelodi, 12 from Ga-Rankuwa and Mabopane and five from Atteridgeville at Pretoria station between 06:00 and 08:00, as well as the arrival of eight commuter trains from Ga-Rankuwa and Mabopane at Belle Omre Station.

06:30 Early morning traffic peaks until 07:00 on the main arterials in the outer city. This includes a centripetal movement of mostly Africans to the buffer zone industries as well as the city core and a centrifugal movement of mostly whites to the buffer zone industries.

06:55 The average worker in Atteridgeville/Laudium and Mamelodi/Eersterus leaves for Pretoria.

07:00 Industrial workers start their shifts.
Within a period of 30 minutes most whites leave home for their daily activities.
Morning traffic peaks until 08:00 on the secondary as well as the main arterials in the inner city. An average of 2,500 vehicles en route to the Pretoria CBD pass within the hour on the main arterials; this includes 2,000 private vehicles and kombi-taxis as well as 50 buses.

07:15 The average African domestic worker reports for duty within the white suburbs of the core city.

07:30   Many Africans have a quick breakfast at the many cooking houses on the fringe of the CBD.

08:00   White-collar professionals have an early morning meeting.

During the morning period the African components of the city are abandoned by resident adults working in the white core area; children are centralized in schools while the infants are under the supervision of mostly elderly people. Conversely there is no change in the age structure within the white suburbs; white adults have only been replaced by African adults. However, there is thus a 'blackening' of the 'white' suburbs as well as the prevalence of women during daytime.

Children are centralized in schools, nurseries or day care facilities.

12:30   Lunchbreak. While whites flock to 'tea rooms' (cafés) and restaurants, Africans crowd in soup kitchens and courtyards. The restrictions of petty apartheid and the lack of alternative facilities inhibit the movement and socializing of blacks and especially in the CBD they resort to public open spaces.

13:00   During the afternoon period white children are either occupied by educational after school activities or under supervision. In the African towns and settlements the children flock to the streets unsupervised. Domestic workers gradually return to their homes in African areas.

16:00   Beginning of the afternoon traffic peak in the inner city.

16:15   Civil servants leave their offices.

16:45   The average white person arrives at home.

17:00   Workers in industries and the private sector leave for home.

17:20   The average worker from Atteridgeville/Laudium and Mamelodi/Eersterus arrives home.

17:30   Late afternoon traffic volumes reach a peak in the outer city.

18:50   The average worker from the border towns of Bophuthatswana arrives home.

19:20   The average worker from KwaNdebele arrives home.

19:50   The average worker from north-eastern Bophuthatswana arrives home.

20:00   Families are united. Many whites use the various public social and recreational facilities within their domain. Africans socialize informally in the towns and settlements.

22:00   As night falls the African, Indian and coloured components of the city are almost exclusively occupied by a particular ethnic group. In the white suburbs up to 20 per cent of the nocturnal population is African, accommodated in rooms or outbuildings.

The totality of these extraordinary daily experiences is indeed a tale of homes apart.

Post-apartheid Pretoria

In presenting a scenario for a post-apartheid Pretoria urban complex it is assumed that all forms of racial discrimination have been removed from the statute books. However, the spatial structures, albeit in a dynamic changing situation will remain functional. One question concerns Bophuthatswana – will it retain its present status of an 'independent' state, or will it flounder? Despite the current resistance of President Mangope, the signs are that it will come under ever increasing pressure to abandon the independence it opted for under the apartheid system. Irrespective of Bophuthatswana's future, the territory's eastern districts and towns will remain an integral part of the Pretoria urban complex.

It is also presupposed that Pretoria will function within a predominantly capitalistic or free enterprise society, where central government will maintain a strong position, though decentralization of power will be actively pursued. Within this context 'normal' urban processes and structures linked largely to the financial capabilities of individuals and groups and to their expertise, will dictate where people live and work. State involvement will be most evident in the provision of land for urban development as well as in financial and supply systems catering for the housing needs of the 'have nots'.

As elsewhere in the country it is predicted that large numbers of people will migrate rapidly to the metropolitan complexes; greater Pretoria's urban nuclei will have swollen to 1.65 million by the end of the century (Table 10.1). To this total the peri-urban population has to be added, pushing the figure probably well beyond two million. The majority of newcomers will, to an overwhelming extent, be the unskilled African and unemployed poor. They will bring great pressure to bear on authorities to make new land rapidly available for housing, and on the provision of infrastructure and employment opportunities; about one person in three is already unemployed. Simultaneously the trend to erect backyard shacks, outbuildings and free-standing shacks will continue. In Mamelodi, on a 300 square metre stand, already four to five shacks accommodating as many as 50 people are part of the urban fabric. A recent study (LHA, 1990) quotes Atteridgeville as having an average of 11.4 people per household, and notes that, of the 99,300 African dwelling units in the Pretoria region, about one-third are shacks and in Mamelodi as many as 48 per cent. Large-scale, high-class urban development characteristic in the past of white residential areas, and even significant middle-class development, irrespective of area, will in future not continue unabated. The greater need will be at the lower income levels where any form of formal housing is largely unaffordable. A predominantly African and poorer Greater Pretoria seems a foregone conclusion. Greater activity will concentrate on the lower end of the housing provision continuum such as spontaneous settlements with minimal services and site-and-service schemes, primarily on the urban outskirts. Their location will be dictated by the present urban structure.

Despite this scenario of ever more densely populated African towns, the creation of urgently needed spontaneous settlement areas and the rise of squatter concentrations where authorities turn a blind eye or permit squatting, residential areas will remain largely ethnically exclusive in terms of tenure. Some Africans, Indians and coloureds with financial ability will move out of existing group areas to what are now up-market white areas, but the numbers will be relatively small. From an economic point of view, it is likely that greater numbers of Africans and coloureds will seek to move into the lower-income white suburbs, especially those closest to the present African or coloured towns. If their entrance into white suburbs is successful, many whites will in turn move out to predominantly middle and higher income areas. Initial racial tension and conflict is likely as people of a perceived lower income group 'invade' higher income white areas. With the passage of time whites could be replaced almost completely in such suburbs. In order to survive in the new urban milieu, overcrowding could arise as accommodation provision is seen as a source of income generation.

To allow lower income people to cater to a greater extent for their own needs, the present high building standards will have to be reduced and unnecessary restrictions removed. The practice of backyard squatting so common in the African areas, though not foreseen in higher income residential areas, will probably find expression in home extensions incorporating additional rooms and apartments. Existing high-rise apartment areas will exert some pull, especially on upwardly mobile younger Indian and coloured couples. The practice of providing high-rise apartment blocks in derelict areas in the vicinity of the CBD or near industrial zones for lower-income families seems somewhat unlikely as this way of living is strange to the majority of South Africans.

Informal activities will no longer be limited to specific areas, but logically will locate where commuter concentrations occur, that is, in the CBD and industrial areas, and also at railway stations and bus terminals. With growing unemployment, crime rates are likely to continue to escalate.

The present transport system will be altered in that the kombi-taxi sector will expand even more dramatically − this form of transport, to an increasing extent, keeps the city functional. Attempts to concentrate embussing and debussing will fail. Mass train transport will serve increasing numbers of commuters, but existing trends will continue. In general, higher income groups will avoid mass transport.

The administrative character of the Pretoria region is likely to be considerably simplified. At the very least, Mamelodi, Atteridgeville, Eersterus and Laudium will be integrated with the present Pretoria municipality into a single city. Verwoerdburg and Akasia might survive as separate municipalities, although their exclusivity will disappear with increasing non-racial urbanization in their areas of jurisdiction. Because of its distance from the core of Pretoria, Soshanguve may continue to function as a separate municipality, but with the probable disappearance of Bophuthatswana as an 'independent' state, integration with Mabopane or even with the wider Winterveld area is possible.

A single authority dealing with planning and environmental services in the whole of the Pretoria functional region is a strong possibility, depending on what initiative is taken at the national level to create such authorities. The extent of such an authority would need to extend far beyond that of the existing Regional Services Council, to include those parts of Bophuthatswana and KwaNdebele which are within the daily commuting hinterland of Pretoria.

Within the whole system, the role of white municipalities and/or authorities

will change drastically. Their stranglehold on urban developme
as the realities of a changing South Africa with its increasi
sector and the accompanying need for jobs, housing and inɪɪas.
superimposed on the old apartheid system. The structures and social patterns
of the 'multiple city' created by apartheid will nonetheless play a significant
role for many years to come in the form and function of the greater Pretoria
urban region.

11
MAFIKENG-MMABATHO
J. H. Drummond and S. Parnell

Within the South African context the town of Mafeking has long been unique. For over 70 years it was the extra-territorial capital of Bechuanaland (present-day Botswana). For a short period from the mid-1960s to the mid-1970s it no longer served as a capital and the town declined, assuming the character of a small country *dorp*. However, with the rise of territorial apartheid a number of showcase bantustan capitals were created. Mafikeng-Mmabatho[1] is the most 'successful' of these. Notwithstanding the wide rejection, and possible abandonment, of the 'independence' of Bophuthatswana, Venda, Ciskei and Transkei, extensive infrastructural development in these bantustans cannot be ignored. The processes associated even with sham independence pose questions of the nature and extent of transformation of urban areas such as Mafikeng-Mmabatho; questions which will have to be faced by the first generation of post-apartheid planners.

This chapter outlines the historical evolution of the town. It then reflects on the theme of Mafikeng-Mmabatho, a bantustan capital, as a city in transition. Questions are posed as to the possible political settlement of the South African crisis and the implications for this bantustan boomtown.

Mafeking as colonial capital

The decision in 1895 to locate the British colonial administrative offices of the Bechuanaland Protectorate outside the colony on crown land in Mafeking known as the Imperial Reserve ensured that Mafeking would always be somewhat different from other South African country towns. Unlike all the other cities considered in this book Mafeking was not initially established by Europeans. In the mid-nineteenth century the Molema section of the Barolong people decided to settle on the banks of the Molopo River. The people settled in small clustered villages which together formed a pre-industrial urban area of some extent (Parsons, 1982). It is this traditional Tswana settlement which is now colloquially known as 'the Stadt'. Following European settlement, racially distinct residential areas with grid-iron street patterns were laid out in the late nineteenth century. The European town was laid out to the north of the Molopo River, on land leased from the Tswana chief, immediately adjacent to the Stadt (Parnell, 1986). The native location and a separate 'half-caste location' were developed on the southern bank of the river. The Imperial Reserve acted as a buffer between the European controlled townships and the tribal settlement of the Stadt (Figure 11.1).

Just why it was ever considered necessary to establish an African location when the Stadt provided an ample supply of labour to the growing town of Mafeking is unclear. The municipal location was controlled by the Mafeking Town Council and the Stadt was administered by tribal authorities who allocated sites for domestic construction and farming. Physical conditions of the two areas were much the same. The 'green' mud brick houses without formal sewerage or piped water in the traditional tribal settlement were also found in the town's location. After 1925 there were more substantial differences between the Stadt and the location. In that year the Natives (Urban Areas) Act was applied to the location, restricting brewing, retailing and African tenure. Theoretically, this legislation should have discouraged African access to urban areas. In Mafeking however, it seems that the location offered Africans easier entry to the town than the Stadt. Migrants, many of whom were Rhodesian, were not eligible for land from the Tswana chief, but if they were employed they were accepted either as lodgers or as standholders in the location. By the late 1940s less than half the household heads in the location were Tswanas.[2]

The African township was administered as a separate residential unit under the Natives (Urban Areas) Act. It was, however, contiguous with the coloured location that fell directly under the jurisdiction of the white council. By the 1930s and 1940s the expansion of the black population had effectively unified the town's locations. This integration was frowned on by the town's officials on the grounds that coloureds, who were permitted to purchase liquor, lived and socialized with Africans for whom alcohol was prohibited. Residential mixing of racial groups was evident in several ways in inter-war Mafeking. Working-class Indians lived south of the Molopo River, while more affluent Indians had houses in the white sector of Mafeking. Even though some sites in the white town had restrictive racial clauses a small number of Indian traders owned property and were included on the city's voters' roll. The co-habitation of Indians and whites seems to have been happily tolerated. More surprisingly, a prominent white attorney and the local doctor lived in the Stadt, a fact that drew little comment at the time. As late as 1950 efforts were made by white lawyers to have the segregationist restrictions on Indian land ownership removed. But, as the entire Indian population never exceeded 200 people, and all coloured and African housing was located across the Molopo river, the colonial character of white Mafeking was hardly challenged.

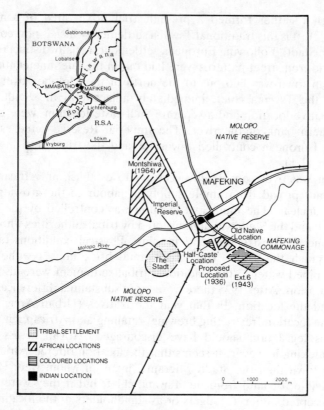

Figure 11.1   Mafikeng prior to 'independence'

Segregation enforced

Residential segregation of the coloured and African communities of Mafeking
was well established by the end of the nineteenth century. In the decades that
followed no attempt to overturn the discriminatory status quo was ever suggested
in Council chambers. Nevertheless, both the African and coloured townships
were forcibly relocated, a trauma normally associated with the implementation
of urban race segregation. In the Mafeking case the reconstruction of the black
locations on new sites was justified in the name of sanitation. *In situ* township
upgrading and expansion were rejected as the solution to overcrowding. Instead,
the Council sought larger sites that would allow for residential expansion
without encroachment on to territory allocated to another race group, or
ground reserved as a buffer strip between races. The local authority's concern
for public health in the white town and their unwillingness to interfere with
the established principle of separate residential facilities for each race group
necessitated forced relocation.

By the mid-1930s the national concern surrounding the sanitation of black
urban locations had reached Mafeking. The Medical Officer of Health expressed
concern over overcrowding and the use of communal latrines in black areas,

noting that improved sanitation in the locations was desirable 'for the sake of both Europeans and urban Natives'. In 1936 a new coloured township was mooted, the idea being that the African location could then expand naturally into the vacated structures, but the proposed site was found to be too stony to build on. A decade later, land due south of the unacceptably rocky ground and a little farther from the town, was proclaimed as 'Mafeking Extension Six, for Coloureds'.

The removal of coloureds to Extension Six, now Danville (Figure 11.2) was a rather slow affair. Stands were on sale from early 1950, but the lack of water connections discouraged building and prevented occupation until the early 1960s. Construction was also delayed by registration difficulties under the 1950 Group Areas Act. Delays in moving the 'half-caste' location, as it continued to be referred to, fuelled suggestions that, as the 'Native Location' was in such disrepair, it should be moved instead. In 1942 the National Health Inspector was forced to report that the appearance of the Mafeking location, 'like so many others (was) not pleasing'. It was now argued that the redevelopment of the African location would allow the coloured area to expand.

The complicated procedure of identifying a suitable site for an enlarged location began in earnest in 1944. Initial suggestions of settling Africans on the Imperial Reserve were quickly dropped as complications of administering a South African township on Bechuanaland soil were realized. Frustrated by the difficulties of acquiring an acceptable piece of ground that was not part of the Native Reserve, that did not belong to another sovereign state and did not abut white suburbs, the Council considered abandoning the removal of the location. For the first time the intervention of the national authority in Mafeking's urban development was evident. Denouncing the scandalous overcrowding of the old location the central government, under Smuts' United Party, quickly sanctioned the exchange of a portion of the Mafeking Commonage for a smaller piece of the Molopo Native Reserve (Figure 11.1) to ensure that the construction of a new African location went ahead as suggested.

By the time the proposed African location finally reached the concrete planning stage, the National Party had assumed power, and was centralizing many matters concerning urban Africans. The plan for 'Montshiwa Bantu Township' was eventually endorsed in 1962 by the central rather than the local authority. Four years later building began. Eviction orders were served on the residents of the old African location and minimal compensation was paid on their structures. Many, but not all, of the African population took 'voluntary' occupation of the matchbox houses that characterized so many of the African townships built in the 1950s and 1960s.

Although residential segregation was being enforced there was, even after the advent of apartheid, still a measure of social integration. As a colonial capital Mafeking had a substantial community of British expatriates and Batswana civil servants, and was perceived by whites in the town to be very different in character, in such things as attitudes to race relations, from neighbouring Afrikaner *dorps* such as Lichtenburg and Zeerust. The Anglican priest at Zeerust noted that in Mafeking mixed congregations 'worshipped together

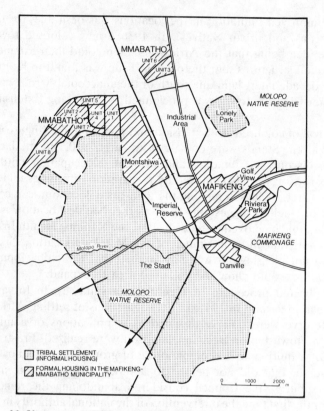

Figure 11.2   Mafikeng post—1977 'independence'

without friction' while such a situation would have been unthinkable in Zeerust (Hooper, 1960, p. 51). As late as 1966 it was possible to argue publicly that Mafeking was 'more cosmopolitan than any other city or town in the Republic. One of its outstanding virtues is that here multi-racialism is an accepted fact. It works without problems, without conscious effort. It is so commonplace that few find it worthy of mention' (*Mafeking Mail and Protectorate Guardian*, 25 February 1966).

The withdrawal of the Bechuanaland Protectorate

The question of relocating the Bechuanaland Protectorate headquarters was mooted even before 1939. Discussions continued throughout the 1950s, culminating in 1961 with the decision to transfer the capital to a site within the Protectorate (Dale, 1969). The loss of this administrative function in 1965 undoubtedly had severe multiplier effects on the Mafeking economy. The town lost a considerable number of jobs, lost its status as a colonial capital and became a relatively obscure *platteland dorp* in the northern Cape. The local authority turned to the central government for assistance, asking to host a new airforce pilot training centre and an army camp in an attempt to attract new inhabitants and wealth to the town. These were refused, causing temporary, but acute hardship for Mafeking (Dale, 1969). The local newspaper noted with alarm the collapse of the local chamber of commerce, and embarked on a campaign for its re-establishment, as well as urging the government to attract border industries to Mafeking (*Mafeking Mail and Protectorate Guardian*, 24 June 1966). But Mafeking was (and still is) too remote from the major markets, and the town failed to attract industrial development. The most significant decision for the future development of the town came with the government's 1963 purchase of the Imperial Reserve for the offices of the Tswana Tribal Authority (*Mafeking Mail and Protectorate Guardian*, 26 April 1963).

As the offices of the Tswana bantustan were woven into the urban fabric of Mafeking, the town was moulded in new ways by the impress of apartheid policies. Perhaps related to Mafeking's loss of colonial capital status, the architects of apartheid took the chance to reshape the town into a more acceptable racial order. Central government intervention and imposition of the draconian Group Areas Act sealed the revised segregationist plan that the Mafeking Council had been planning for the past two decades. In 1963 the old coloured location was rezoned for whites, although it appears that this was never enforced, as to this day several coloured families continue to live in the original houses. Indians, who had lived amongst both black and white, were moved with only minimal protest into a single group area in 1969. Two years later, the original African location, by then cleared of the 15 families who had strenuously resisted moving to Montshiwa, was officially proclaimed the coloured group area of Danville.

At the time of Bophuthatswana's 'independence' in 1977 Mafeking had thus been shaped by apartheid ideals, being composed of the white and Indian areas of Mafeking, the coloured area of Danville and the African areas of Montshiwa and the Stadt (Figure 11.2). As a result of 'independence' several transformations were wrought in the urban fabric of Mafeking.

Post-'independence' population growth and housing markets

Mafeking's period of decline as a *platteland dorp* ended in 1976. According to Cowley (1985a) Mafeking had tried hard to persuade the government of Bophuthatswana to locate the capital in the area. In December 1976 the Bophuthatswana legislature decided to establish a 'capital city', called Mmabatho, 7 kilometres north-west of Mafeking. 'Independence' in 1977 triggered an economic boom for Mafeking (Parnell, 1986). The major impact of new construction was felt in Mmabatho but the spin-off effects were experienced in Mafeking which, despite being outside Bophuthatswana, initially served as the commercial centre for Mmabatho (*Mafikeng Mail and Guardian*, 18 September 1980). The immediate effect of developments in Mmabatho was the generation of substantial new job opportunities in construction, in tourism and in the civil service. Associated with this was a growing housing shortage. Confirmation that Mafeking would continue to participate fully in the economic boom came with the decision in 1980 to incorporate the town into Bophuthat-swana. This event was warmly welcomed by the business community of Mafikeng, as it was henceforth known (*The Star*, 27 March 1980). For some of the older inhabitants, the recovery of 'capital city' status was more than welcome.

Since 1977 the physical size of the Mafikeng area has expanded dramatically (Figure 11.2). The population grew from roughly 35,000 in 1976 to over 85,000 in 1985 (Cowley, 1985b). Current estimates for the contiguous urban area are 140,000 (Cowley, 1990). It should be noted that several resettlement villages, accommodating Africans relocated from white farms, exist within a 25 kilometre radius of Mmabatho (Surplus People Project, 1983; Parnell, 1986). If these settlements are included, the population of the functional city region is more likely to approach 170,000. This figure has been used by planners for feasibility studies for Mafikeng-Mmabatho services (*Mafikeng Mail*, 17 October 1986). Three distinct groups of in-migrants can be identified: first, both African and white administrators, civil servants, Third World expatriate professionals and educators associated with state and quasi-state functions; secondly, an influx of African families resettled in the area; and thirdly a disparate stream of African voluntary migrants.

In considering the effects of population growth upon the urban fabric of Mafikeng, it is important to recognize the manner in which employment is linked to residential access. First, differentiation occurs between those in formal employment and those unemployed or underemployed. Secondly, of those employed, opportunities for shelter differ considerably between state and private sector employees. Since 'independence' race has been de-emphasized as the exclusive determinant of residential access differentiation. New categories of residential housing have emerged in line with the shifting economic structure of Mafikeng.

The most immediate changes after 'independence' and incorporation involved the abolition of statutory residential segregation by race, and the introduction of controls on non-Tswana land ownership, whereby the permission of the Minister of Urban Affairs had to be obtained for house purchases (Section 12, Act 39 of 1979). This legislation has been selectively applied to control speculation in the housing market. Despite the need for ministerial consent and the restriction of one property for each owner, the town's middle to upper-income property market boomed after 1977. The previously all-white housing market of Mafikeng was relatively unaffected, partly due to the construction of the new élite, racially mixed, freehold suburbs of Riviera Park and Golf View (Figure 11.2).

The fact that the Mafikeng housing market has moved through the transition to 'independence' without any fundamental changes in composition may be attributed to the high cost of houses in the area, which were maintained by subsidies from the South African Development Trust. There has been no major influx of élite Africans, with blacks accounting for only about one in six residents of the former white areas of Mafikeng (Pickard-Cambridge, 1988b).

There has not been a pronounced exodus of whites from Mafikeng, partly because the Bophuthatswana authorities in 1980 allowed the white schools and hospital to remain white for five years, subsequently extended to 10 years, to smooth the transition to a non-racial town. Louw (1986) sampled a group of white property owners in Mafikeng, the majority of whom described race relations as good to excellent, but also indicated they would leave the town if schools ceased to remain for whites only. Referring to Mmabatho, Pickard-Cambridge (1988b, p. 45) suggests that 'the transition to a living environment desegregated both formally and in practice' has been made, citing as evidence the frictionless settlement of upper-income suburbs of Mmabatho. Although the prestige suburbs of Riviera Park and Golf View have acquired a more obviously white upper middle-class character than Mmabatho, substantial numbers of black families have built in these exclusive new suburbs.

The lower levels of the Mafikeng-Mmabatho housing market present a more complex process of transition. The most obvious cleavage exists between the accommodation available to those in employment on the one hand, and to those unemployed or underemployed on the other. Whereas the former group enjoy access to a range of 'formal' housing, the latter must generally have recourse to 'informal housing', usually on tribal land. Within the category of formal housing a second distinction emerges in the post-'independence' period between those who qualify for state housing and those who do not.

Prior to 'independence' influx control restrictions effectively divided the two African residential areas of Montshiwa and the Stadt. As only those persons with Section 10 rights officially qualified for employment in Mafeking, it follows that Montshiwa township legally offered accommodation only for those drawing a regular income. Those lacking Section 10 urban rights lived in the 'informal' settlement of the Stadt. With 'independence' and withdrawal of influx control, the situation changed only marginally. The unemployed and poorest people still could not afford the costs of formal housing. What did change, however, was the size of the population forced to live on tribal land. Associated with the economic boom there has been an influx of new migrants looking for unskilled work in the new capital. These people find accommodation on tribal land where, for a fee paid to the local chief, they are able to build their own shacks. One consequence has been the extension of the Stadt (Legge, 1984), and the creation of new residential areas of predominantly informal housing (Figure 11.2). An important aspect of this transformation is the emerging conflict between chiefs and government officials who view these areas of 'informal' housing as an eyesore in the developing capital (Van Wyk, 1984).

More immediately obvious than the expansion of the Stadt and other tribal areas has been the provision of over 6,000 formal housing units in Mmabatho. Built primarily for state employees who may choose either to purchase or to rent, these units are not easily obtained on the open market. The hierarchical organization of the civil service and other state organizations is already clearly apparent within the residential area of Mmabatho. Units three and five house largely the new racially diverse élite. At the other end of the scale units seven and eight correspond in status to 'economic housing' of other South African cities (Van Wyk, 1984). The impact of state dominance on the urban fabric is reinforced because of the lack of alternative accommodation in the area.

Apart from the more expensive Mafikeng houses the only other lower to middle-income housing available is that of Montshiwa (Figure 11.2), where the dominant form of tenure is now home ownership (Moadira, 1984). The records of the property register reflect that the move to purchase these houses occurred after 'independence'. Two possible explanations of this transition to home ownership may be offered. First, as only existing township tenants were able to purchase their homes (Van Wyk, 1984), the social and economic status of these residents must have improved after 'independence'. The second interpretation is that tenants bought their homes in order to sell at an inflated price to those unable to find accommodation in Mmabatho, and have moved to the Stadt or some other cheaper accommodation. What emerges unequivocally is that in Montshiwa the tenural relationships in the township have undergone a substantial transformation since independence (Moadira, 1984).

It is evident from this discussion that there has been a breakdown of residential segregation. Specifically, there has been a small movement of middle-class Africans, coloureds and Indians into Mafikeng's formerly exclusive white suburbs, and a more marked trend towards residential integration in the new élite areas of Mmabatho units three and five and in the upper income freehold areas of Riviera Park and Golf View. The expansion of the Stadt and other informal housing areas reflects a post-'independence' process of rural—urban migration typical of most African and Third World contexts. The immigration of white South Africans as well as a significant number of expatriates into the Mafikeng area is paralleled in cities such as Harare and Windhoek where a community of expatriates concerned with foreign aid and development issues has settled, at least temporarily.

Since 'independence' there has been a radical transformation in the town's economic base from reliance on the provision of agricultural and transport services to domination by state and quasi-state functions. One area of private sector growth has been tourism, fostered by the Bophuthatswana government's concessions to gambling. Associated with the economic transformation in the capital has been the emergence of construction as a vital, though probably ephemeral component of the area's employment. Linked to this metamorphosis of the urban economic base has been the evolution of a residential landscape which is no longer entirely dependent on race, nor wholly linked to economic status. The dominant theme which emerges in the new residential market is one of differentiation along the cleavage of access to formal, more particularly, state employment. The critical significance of the state as the provider of basic housing emerges as a factor separating those who are civil servants from private sector employees. To some extent therefore the property market in Mafikeng is very atypical. The demand for houses for state-linked employees has led to a buoyant and over-inflated housing market whose prices are comparable with Johannesburg rather than other South African country towns.

Mafikeng-Mmabatho in a post-apartheid South Africa

At the time of writing Bophuthatswana remains the only bantustan which is committed to preserving its apartheid-conferred 'independence'. However, this path seems doomed to failure as the African National Congress campaigns for a unitary, non-racial democratic South Africa and the South African government has indicated its opposition to continuing with the structures created by 'grand apartheid'. In this political climate the question of the effect of the re-incorporation of Bophuthatswana into South Africa is of immediate interest.

Although it is widely accepted that Bophuthatswana will collapse, its inevitable demise is being most vociferously resisted in Mafikeng-Mmabatho, the region which has benefited most from 'independence'. Ironically many of the post-apartheid changes most feared elsewhere in South Africa have already been successfully overcome in Mafikeng-Mmabatho. Thus far, residential desegregation has not undercut property prices and even conservative whites express satisfaction with race relations in the town (Louw, 1986). The critical question for the future is whether the urban area can sustain the prosperity that has fostered the harmonious climate of the past decade once the town is deprived of its artificial status of bantustan capital.

Since 'independence' in 1977 the Mafikeng-Mmabatho economy has flourished. However, there is no productive base for continued economic prosperity in the area. It is unlikely that the agricultural processing industry can be developed much further. If re-incorporation were to rob Mafikeng-Mmabatho of its function as a capital city the town would, as occurred in the mid-1960s when Gaborone was made the new capital of Botswana, face severe hardship. It is likely that one immediate effect would be a collapse in property prices, as there would be doubts cast over the continued existence of the para-statal and state bureaucracies which have been the source of growth in the housing market. Although many home owners would be affected by the loss of state jobs, unskilled and semi-skilled workers living in the less prosperous parts of Mmabatho and Montshiwa are least likely to be able to find alternative employment or meet their repayments.

Unless a post-apartheid government were to establish a regional capital that would allow for the transferral of personnel to revised administrative tasks, a flight of skilled labour is inevitable. As it is unlikely that a future government would bolster house prices of the middle-class, as the South African government did through the subsidies of the South African Development Trust, this section of the property market of Mafikeng-Mmabatho is likely to face more severe challenges than those of the early 1980s when Mafikeng was incorporated into Bophuthatswana and desegregation threatened white housing prices.

In addition to the undercutting of the property market, the commercial base of the town would be adversely affected by re-incorporation. Although the population of Mafikeng-Mmabatho has increased dramatically, the predominantly poor population is unable to sustain the extensive commercial developments that now exist. The shopping complexes in both Mafikeng and Mmabatho

flourish because of the patronage of neighbouring towns like Zeerust, Lichtenburg, Vryburg, Lobatse (Botswana) and even Gaborone. The attraction of
shopping in Mafikeng-Mmabatho is that Bophuthatswana does not levy the
13 per cent general sales tax common to all South African shops. As re-
incorporation would remove this relative advantage, the likelihood of Mafikeng-
Mmabatho retaining its current status as a regional service centre is uncertain.
Similarly, its attraction as a tourist centre would be undermined if subject to
the same gambling laws (permissive or otherwise) as the rest of South Africa.

Other sectors will also be adversely affected by incorporation. The construction
boom has been almost exclusively founded on building either for government
or for government employees, and some decline is inevitable even without the
demise of an 'independent' Bophuthatswana. There are many other unskilled
workers, particularly in domestic service, who would be unemployed without
the numerous state personnel currently located in Mafikeng-Mmabatho. In
short, the likely economic scenario of re-incorporation does not augur well for
many of the town's people.

Given the extensive infrastructure that now exists in the Mafikeng-Mmabatho
area it would seem reasonable to assume that a new South African government
would give serious attention to the future role the city could perform in a post-
apartheid era. The most obvious function for the town would be as a regional
capital for the western Transvaal and northern Cape area. There would be
clear advantages to adopting such a proposal, not least being the more responsible utilization of the South African government's considerable investment in
the area. As a regional capital in a post-apartheid South Africa, it is probable
that relations with Botswana would be normalized and trade links strengthened
(Drummond and Manson, 1990).

If a new government were to decline to reconstitute the present-day bantustan
capital of Mafikeng-Mmabatho as a new regional or provincial capital, and
indeed there may be political reasons for this, there are other potential avenues
for preserving the local economy. The town already boasts an unusually large
number of educational institutions, including a university. Given the relative
undersupply of tertiary training centres in South Africa, Mmabatho could
profitably be developed as an educational centre. A second option for the
town's economic preservation would be for it to become a military centre.
Aside from its locational advantages on the country's western border, there is
an army base of some size on the northern perimeter of Mmabatho. In addition
to a nearby South African radar station, the Mmabatho international airport
already offers the most modern facilities and a second runway for the
Bophuthatswana Airforce is currently under construction. Clearly, post-
apartheid planners need to give urgent consideration to the future of Mafikeng-
Mmabatho and other bantustan capitals.

Dismantling apartheid is not an easy task, but the problems to be faced in
building a new South Africa in Mafikeng-Mmabatho differ from the other
major urban centres, although no doubt similar issues will be identified in
Umtata, Bisho, and Thohoyandou. Mafikeng-Mmabatho has shown that the
desegregation of residential areas need not encourage racial hostility. The

atmosphere of racial tolerance since 'independence' has been fostered by the presence of refugees from South Africa's Mixed Marriages Act and military conscription, and by the existence of a cosmopolitan community of Third World expatriates. The guarantees given to Mafikeng's white community may also have contributed to relaxed race relations. A further explanation is that the destruction of apartheid urban practices was not challenged by white and black Mafikeng-Mmabatho residents because their personal livelihood was enhanced, rather than threatened, by living in the 'capital city' of Bophuthatswana. Similar support for a new post-apartheid government will only be secured if this economic stability can be assured.

Acknowledgement

Thanks are due to Chris Rogerson for comments on an earlier draft of this chapter.

Notes

1. Mafeking is the Anglicized version of Mafikeng – the original Tswana place name. With the incorporation of the town into Bophuthatswana in 1980, the name reverted to Mafikeng. Mmabatho is the name given to the 'capital' city of Bophuthatswana, which was developed a few kilometres north-west of the colonial town. In 1984 the two municipalities were united and presently the city is known as Mmabatho, of which Mafikeng is a suburb. Up to 1980 the term Mafeking is used; thereafter Mafikeng refers to the incorporated town. The greater settlement is known as Mafikeng-Mmabatho.

2. Archival material for this and the following section of the paper is housed in the Cape Archives Depot. The series 1/MFK and 3/MFK provided particularly rich information. Detailed references may be obtained from the authors.

12
WINDHOEK: DESEGREGATION AND CHANGE IN THE CAPITAL OF SOUTH AFRICA'S ERSTWHILE COLONY

David Simon

The primary objectives of this chapter are to analyse the nature and consequences of urban desegregation in Windhoek since political independence for Namibia first became a realistic possibility in 1977, and to evaluate the possible lessons for urban change in South Africa as the post-apartheid era draws steadily closer. However, it is important to understand these recent developments in historical context, and the first sections trace the origins and evolution of colonial urbanism in Windhoek, focusing on the implementation of segregation and apartheid policies so as to inform the subsequent discussion. More detailed historical coverage can be found in Simon (1983).

Windhoek's significance for South African urban futures lies not only in the geographical contiguity of Namibia and the recency of desegregation there, but also in the fact that, under South African rule since 1915, Windhoek had become a model apartheid city. Namibia's eventual independence on 21 March 1990 was the culmination of a transition period lasting over 12 years during which South Africa defied the international community while trying to engineer a neocolonial solution. This involved delaying independence on various pretexts while continuing to hold the line militarily against SWAPO guerrillas, despite the high human, financial and political cost. This provided the time and space to install and nurture successive versions of a 'moderate' and pro-capitalist client regime which it hoped would be capable of defeating SWAPO in the UN-supervised pre-independence elections. It was as part of this process that the dismantling of much overt apartheid legislation occurred. The South African state observed the changes closely, regarding Namibia as something of a social laboratory in this respect, especially in terms of white reactions to desegregation.

More generally, little comparable work to the research summarized here and in Chapter 13 has been undertaken in cities elsewhere in the Third World during the decolonization process, and it has therefore also contributed to theoretical conceptualizations of the urbanization process (cf. McGee, 1971; King, 1976, 1985; Simon, 1984a, 1989b).

From kraal to colonial capital

Archaeological evidence reveals that human activity on the site of Windhoek dates back at least 5,000 years, due, at least in part, to the perennial springs which sustained life in a semi-arid environment. A succession of indigenous groups has occupied the site, most recently the Orlams and Herero, but the formal town dates from the arrival of German colonial settlers in 1890. The settlement quickly became the capital of the German settlement colony of Deutsch Südwestafrika, occupying a position of disproportionate socio-political importance despite its small size. In contrast to other settlements in the territory, 'a more discriminating social life grew up in Windhoek based on the social hierarchy of a German provincial town' (Bley, 1971, p. 77). Class distinctions among the settlers were thus important *ab initio*. Labour was provided almost exclusively by subjugated and forcibly resettled Africans. In the wake of further conquests and land dispossession to provide farms for new German settlers, more Africans were forced by poverty and the lack of alternatives to survive by selling their labour in the town. Since Germany established an empire far later than its European rivals, its officials modelled their 'native policy' on precedent, particularly that of the British in the neighbouring colonies which comprise present day South Africa. When the Germans surrendered to South African forces in 1915, Windhoek's population had reached roughly 7,500 (some 4,500 Africans, about 3,000 Europeans and a handful of people of mixed race [Pendleton, 1974, p. 31; Simon, 1983, p. 91]).

Inevitably, the social use of space at both the domestic and urban scales reflected the dominant value system. The layout of the settlement and the subsequent town planning code developed to formalize and guide growth in the urban fabric were imbued with German colonial norms and values. Their influence is still clearly evident today, in terms both of the architectural styles of the many surviving private and public German buildings, and, more fundamentally also in terms of the urban layout and road network of the central city and adjacent suburbs.

Class distinctions among Europeans found expression in residential differentiation. The imposing homes of the emerging élite spread over the hillside overlooking the town and into the adjacent valley of Klein Windhoek. This assured a more pleasant view and microclimate, while being situated as far as possible from the commoners and African workers. The process was systematized, as in other European colonial territories, through the establishment of private property rights under a capitalist land market based on individual freehold and leasehold tenure.

Racial segregation, a virtually universal characteristic of colonial cities, was also quickly instituted. Small groups of huts on the fringe of the new settlement are evident on the first maps drawn of Windhoek in the early 1890s. Over the years, relocation and consolidation into two controllable 'locations' occurred

(Pendleton, 1974; Simon, 1983). In keeping with overall 'native' policy, Africans were prevented from owning urban property, thus being reduced to occupying tied accommodation of appallingly low standard, or renting plots (and later, under South African control, dwellings) from local authorities. Initially the only formal education for Africans was provided by the missions, but even this was segregated. The first mission schools in Windhoek date from 1894. The state gradually sought to increase control over education, but highly discriminatory provision and financial allocations were instituted from the very beginning (O'Callaghan, 1977a; Melber, 1979). The same is true of health services.

Segregation and apartheid policies under South African rule (1915–77)

In 1919, Namibia became a C-Class Mandated Territory, known as South West Africa, under the League of Nations' system. South Africa assumed responsibility on Britain's behalf and systematically enhanced its control, quickly replacing German legislation. Closer integration and eventual incorporation of Namibia was pursued with greater or lesser vigour until the mid-1970s.

South African economic penetration and white immigration began immediately. The pace of urban growth fluctuated in accordance with cyclical economic conditions, but accelerated after the Second World War as a result of major new South African investments in Namibia and improved agricultural prospects. Windhoek's population grew from just over 10,000 in 1936 (with whites apparently just outnumbering Africans) to 15,000 in 1946, 20,600 in 1951, 36,000 in 1960, 51,000 in 1968, 62,000 in 1970 and 74,500 in 1975. According to these official figures, the number of Africans overtook that of whites only in 1975, but almost certain underenumeration suggests that this in fact occurred some time earlier.

From the point of view of the black (predominantly African) underclasses, the advent of South African rule brought little change. Over time, the scope of discriminatory legislation was increased, usually through the extension of laws closely resembling their South African equivalents. Much of this legislation was national, rather than specifically urban, in nature, governing matters such as 'native administration' and prohibitions on access to credit. The centrepiece of urban segregation policy was the Natives (Urban Areas) Proclamation, No. 56 of 1951. Based on the 1945 Consolidation Act of the same name in South Africa, it superseded and systematized earlier legislation enforcing the pass laws and confining Africans to discrete 'native locations'. Unlike its South African equivalent, however, it was promulgated only after the National Party's assumption of power in 1948. Thereafter, virtually all major apartheid legislation was applied to Namibia, with the singal exception of the Group Areas Act. Its provisions were deemed too sensitive in the context of the territory's disputed status, while the very high customary level of segregation, and the existence of other measures rendered further legislation largely superfluous (Simon, 1986). Other than the Natives (Urban Areas) Proclamation in relation to Africans, the most important segregative device was the inclusion of racially restrictive clauses in the title deeds of individual properties, a practice systematized by the municipality during the 1960s and 1970s. However, some older properties in the central area and Klein Windhoek were never subject to such clauses.

The Main Location to the west of the city centre was formalized on a grid-iron street layout in 1932, and some standpipes and pit latrines provided. Inhabitants paid the municipality a ground rent and erected their own huts, shacks or modest houses. Plans for the construction of a new township to replace the Main Location, first mooted during the 1940s, were revived and acted upon during the 1950s. Although conditions had undoubtedly deteriorated over time and the population had continued to grow, the authorities' principal concern was that the Location was restricting westward expansion of the white city, and such close proximity of such a large African population to white suburbs was deemed undesirable. Apartheid urban planning provided the rationale and the physical designs, and separate new townships for Africans (Katutura) and coloureds (Khomasdal) were subsequently built some 5–6 kilometres north-west of the city centre. Statutory buffer strips separated them from each other and from the white city (Figure 12.1). Katutura contained a large, prison-like hostel for Ovambo contract workers. Unlike Africans, coloureds were permitted to own land and property; however, only a few people could afford this option and some 90 per cent of houses in Khomasdal were rented from the municipality.

Forcible relocation of the population from the three locations occurred, notwithstanding a concerted resistance campaign which began in 1959 and continued periodically until final destruction of the locations in 1968. Henceforth the Main Location became known as the Old Location. The Administrator's refusal to see the protesters triggered a boycott during which the police confronted a peaceful protest meeting in the Main Location on 10 December 1959, eventually opening fire, killing 11 and wounding over 40. This massacre, which predated Sharpeville by three months, has assumed an equivalent place in Namibian history, and served as a spur to the nationalist movement for independence (Simon, 1988a, 1988b). By the mid-1970s therefore, when Angolan and Mozambican independence finally convinced the South African state that the winds of change *were* blowing through southern Africa and precipitated attempts to negotiate Namibian independence on terms acceptable to Pretoria, Windhoek had complied fully with the apartheid city model for some years (Figure 12.1).

Unravelling the apartheid web (1977–90)

The city has continued to grow steadily throughout this period. Projections for 1987 based on the most recent municipal survey estimated the total population at 103,000, comprising 35,000 in the former white area, 53,000 in Katutura and 15,000 in Khomasdal (Municipality of Windhoek, 1987; but cf. Von Garnier, 1986). The author's own estimate for the entire city was about 120,000 in 1986, and probably 140,000–150,000 at independence.

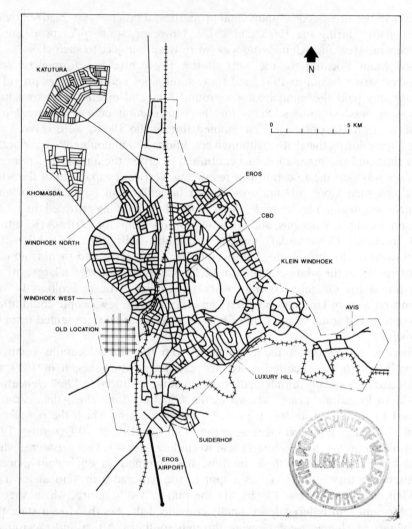

Figure 12.1   Windhoek: the apartheid city of the late 1960s (source: Simon, 1986).

The repeal of discriminatory legislation

Initially it was South Africa's newly appointed Administrator General who
commenced the amendment and repeal of overtly discriminatory legislation by
proclamation in late 1977. Only those measures relevant to urban areas are
discussed here. The first laws to go were the prohibitions on interracial sex and
marriage, most of the Native Administration Proclamation, the Prohibition of
Credit to Natives Proclamation, and the most onerous sections of the Natives
(Urban Areas) Proclamation, including the burden of proof, 72-hour clause,
influx control and curfew provisions. Local authority powers under the Pro-

clamation were reduced while other amendments enabled Africans to purchase land and property in townships for the first time.

In contrast to this single step, made possible by the autocratic nature of the Administrator's rule by decree, freehold rights in South Africa were introduced only in 1986 as the culmination of a tortuous decade-long process of progressively lengthened leasehold systems and legal semantics (Simon, 1989b). The above measures removed some of the most oppressive and hated restrictions on Africans, and sought to bring property and related laws pertaining to African townships into line with those applicable elsewhere in the urban areas.

Formal intra-urban racial zoning, however, remained on the statute books for two more years, until the first National Assembly constituted by South Africa to promote an 'internal solution' passed the Abolishment of Racial Discrimination (Urban Residential Areas and Public Amenities) Act, No. 3 of 1979. This lifted all racially exclusive provisions in residential areas and many public amenities, although the clauses on public amenities and penalties were deferred for a year to appease right-wing whites. In 1981, the scope of the Act was widened to embrace all urban land, while procedures for tracing offenders were improved. Two further amendments to the Natives (Urban Areas) Proc-lamation in 1980 enabled local authorities to levy property rates on freeholds in African townships, and removed the longstanding prohibition on the purchase of property in townships by whites. Since 1981, then, Windhoek has been an 'open' city at both the intra-urban and inter-urban scales, in terms of formal legislation. In theory, people have been able to occupy and purchase freehold title anywhere, while no significant restrictions on rural–urban or inter-urban migration exist. Reality has been somewhat more complex, however.

Quite apart from the attitudinal factors discussed below, several structural and institutional constraints have continued to operate. First, only a limited proportion of coloured people and a very small minority of Africans, could afford the far higher rents and prices prevailing in former white Windhoek. Secondly, some 38 per cent of the urban housing stock was owned by the state and parastatal corporations before sale to employees commenced in the early 1980s. Employers' housing policies therefore had a significant impact on initial mobility. Even where no overt discrimination persisted, the fact that whites continue to dominate the skilled job categories for which good quality housing is reserved, restricted mobility in this sphere for the first few years. More recently, though, much of this housing has been sold to employees, while the number of black politicians and civil servants increased during the 1980s as a result of the proliferation of ethnic second tier authorities, and the gradual takeover of increasing central government responsibilities from Pretoria. Coupled with progressive Namibianization of the civil service, the reduction or elimination of racial wage differentials and the equalization of employee housing scheme conditions, this has increased the demand by blacks for high quality housing in former white suburbs (see also Pickard-Cambridge, 1988b). Thirdly, movement into Windhoek for Africans certainly became far easier once the pass laws were abolished, but various other restrictions remained in force, e.g.

curfew provisions, road blocks and intimidation relating to the war in the North, whence a large proportion of migrants come, and the need to register employment contracts.

Given the wider political context set out earlier, it is important to point out that the objective behind desegregation and the repeal of other discriminatory laws was not to introduce a fundamentally different socio-political order but in fact to remove overt apartheid so as to defuse pressure for more radical change. In other words, the authors of reform ultimately sought to retain the existing, essentially capitalist and statist, system as far as possible while giving it a more human face. During the mid to late 1980s, significant economic deregulation occurred but privatization had scarcely begun by independence.

Attitudes to change

As already explained, desegregation measures hardly affected the lives of most black inhabitants of Windhoek in a direct sense. For many, particularly supporters of SWAPO, they appeared little more than cosmetic gestures, precisely because the underlying structural inequalities were unaffected. Most so-called 'moderate' politicians and their followers and some elements of the fledgeling black commercial petit bourgeoisie welcomed the measures as evidence of progress, though some clearly felt that they did not go far enough. The majority of Windhoek's conservative white population opposed desegregation and related changes quite vociferously when the measures were first mooted and while they were being enacted. This was articulated through elected white representatives at municipal and national levels, where the National Party enjoyed greater support than the Republican Party (the white constituent of the DTA ethnic alliance).

Municipal attitudes at the time can be examined at two levels. First, the municipality sought resolutely to enforce existing legislation during the period 1977–79, when black people first began to filter into white areas in anticipation of desegregation. Complaints from white residents were followed up by the inspectorate, and it was gradually realized that the scale of the phenomenon was far greater than generally believed. Protracted debate therefore ensued between councillors and various departments as to the most appropriate course of action. The position was further complicated from their point of view by contradictions created by the initial legal changes in late 1977, with the result, for example, that people classified as being of different races were already entitled to cohabit or marry, yet not to live in the same dwelling or suburb. Some councillors and professional staff, such as town planners, were also not opposed to desegregation, possibly because, unlike blue collar workers, it posed no threat to their personal or class interests. Moreover, prospects for an international settlement over Namibia at the time made many councillors wary of taking such controversial steps. The upshot of this indecision and procrastination was that little concrete action in fact occurred (Simon, 1986, pp. 294–6).

At the second, public, level the City Council maintained a posture opposing desegregation. This was based on the now familiar arguments about 'forced integration' and the conflict it would engender; the deleterious effect large numbers of 'black and brown' people would have on white areas in terms of 'overcrowding', hygiene and existing standards; and sometimes unveiled racism. These mirrored the attitudes of white settlers faced with decolonization elsewhere in Africa, and are now commonplace in South Africa too. Successive Council debates on the subject were highly emotional, and strong representations were made to higher levels of authority, both directly and through the all-white Municipal Association. Once legal desegregation became inevitable, attention focused on ways to minimize its impact on whites, primarily by attempts to hive off Katutura and Khomasdal as separate municipalities, and by improving facilities there so that moving to former white areas would become less attractive (Simon, 1985, 1986). Over time, however, public attitudes softened as *de facto* integration lost its novelty and the anticipated fall in property values did not materialize (see below). Local authorities, parastatals and other employers owning or administering housing stocks had to abide by the law.

Residential mobility after the abolition of racial zoning

The actual experience of intra-urban residential mobility has been mixed. As stated above, the process of movement into previously forbidden areas anticipated the legal changes. In the main, this involved coloured people moving out of Khomasdal and into flats, rented houses and servants' quarters in and around the central business district (CBD) where discriminatory title deed restrictions often did not apply, although cases were also recorded in a number of other areas. These areas were the equivalent of 'grey areas' in South African cities. Isolated cases of house purchase through white nominees were also reported.

From August 1979, the now legal process continued at a relatively slow but steady rate over the first few years, and accelerated slightly during the mid to late 1980s. The spatial distribution of moves has been very uneven, with the CBD and vicinity continuing to form the single most important destination area, followed by Windhoek West and Klein Windhoek, with smaller numbers in and around the Southern Industrial Area, Luxury Hill and Eros, and a sprinkling in most other areas (Figure 12.2). There has also been significant movement between Khomasdal and Katutura, coupled with a surprising degree of return movement to Khomasdal from former white areas. This pattern is determined by a combination of demand and supply factors, especially the cost, availability and type of accommodation, including the extent of public ownership, availability of loan capital, availability of transport and proximity to the workplace and schools (which, apart from private schools, remained segregated until independence), perceptions of the social character of different areas and the likely reception incomers would receive (see Simon, 1986). For example, parts of the CBD fringe contain the least desirable and hence cheapest housing formerly reserved for whites, who have, in the main, suburbanized as their real incomes rose. Windhoek West has an intermediate and mixed character, while

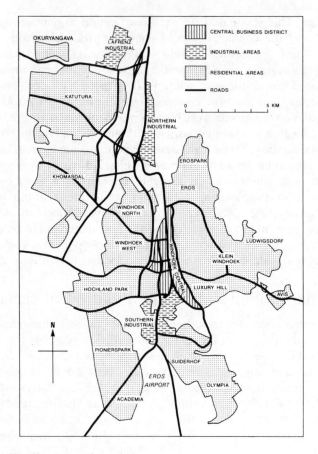

Figure 12.2   Windhoek at independence

lying adjacent to Khomasdal. Klein Windhoek is a large and established middle to upper middle-class suburb with a high proportion of German- and English-speaking whites. By contrast, suburbs such as Windhoek North, Pionierspark, and Academia have high concentrations of tied accommodation, are lower middle to middle-class and predominantly inhabited by Afrikaans-speaking whites. Notwithstanding somewhat lower prices, these have consequently been among the least attractive to black people, although Windhoek North today has a significant black population as a result of the desegregation and sale of houses by the transport parastatal corporation.

The people moving into formerly prohibited areas have been overwhelmingly middle-class or incipient middle-class, under 45 years of age and living either singly or in nuclear families. People classified as 'coloured' have consistently dominated, although the proportion of Africans has risen since the mid-1980s. Numerous other households comprised racially mixed marriages or cohabitations, living, with only one or two exceptions, in former white areas. Very few Africans moved out of Katutura. Two other significant categories of incomers

emerged during the early years, namely exiles returning from abroad in anticipation of independence in the late 1970s, and a sizeable number of coloured
families and mixed couples emigrating from South Africa to escape apartheid
(Simon, 1986). Such migration flows have declined markedly since the reforms
of the mid-1980s in South Africa, especially repeal of the Mixed Marriages Act
and Section 16 of the Immorality Act.

Municipal survey data suggest an overall black occupancy of residential units
in former white areas of just over 12 per cent in 1985. This figure omits
domestic servants and other categories, so the overall black proportion of these
areas' population was slightly higher, at 12.74 per cent. The figures for individual
areas ranged from roughly 28 per cent and 22 per cent in Windhoek Central
and Windhoek West respectively to five per cent or under in suburbs such
as Olympia, Academia and Pionierspark (Municipality of Windhoek, 1987).
Most unofficial sources concur with these orders of magnitude (Pickard-
Cambridge, 1988b). Although the figures will have increased by independence,
it is significant that nowhere are blacks yet even remotely approaching a
majority. Conversely, and as might be expected, well under one per cent of
dwellings in Katutura and Khomasdal are occupied by whites.

The experience of incomers has been varied but generally quite positive.
Especially during the early years, black people were not uncommonly denied
tenancies by white property owners once their skin colour was discovered.
However, after moving, only a small minority of incomers complained about
hostility or harassment by their new neighbours. The most serious examples
involved harassment of children playing in the street, and verbal abuse from
neighbours in certain blocks of flats, but the atmosphere usually improved over
time. In the vast majority of cases, there was either virtually no contact with
neighbours (a common feature of middle class areas internationally) or relations
were reported as being friendly. In general terms, mixed couples suffered most
harassment, and even physical attack in public places, during the first two to
three years. This was perhaps predictable, as to reactionary whites intermarriage
must have been perceived as a serious threat to their concept of racial purity.
A few street attacks on blacks in white areas were reported, but all overtly
racist attacks had declined steadily by the early 1980s as the new order became
'normal'.

The townships: immobility and overcrowding

Longstanding chronic housing shortages in Katutura and Khomasdal have their
origin in the almost complete lack of new construction between 1968 and 1975.
This was a deliberate policy measure designed to minimize the attractiveness of
urbanization. As in South Africa, the prohibitions on African property ownership
meant that all land in African townships was owned by the local authority,
which constructed and let housing as agent of the central state. Construction of
somewhat better quality housing resumed in 1975, but even accelerated rates of
production failed to arrest the official backlog. Official data were based on
municipal waiting lists, which significantly underestimated the true situation.

Average occupation rates in the standard township houses with floorspace of 56 square metres or less in Katutura and 72 square metres in Khomasdal, reached 12 and 18 respectively in 1981, with cases of up to 30 verified (Simon, 1988a). Such data are not, however, universally accepted; official sources tend to show rather lower figures. Hence, for example, the 1985 municipal survey suggested that the mean number of people per dwelling in Khomasdal had fallen to 5.7 from a 1975 figure of 7.6, but a 1986 church survey yielded higher densities both here and in Katutura (cf. Von Garnier, 1986; Municipality of Windhoek, 1987, pp. 172–173). For Africans, the opening of cities via abolition of the pass laws was an important facilitatory change. On top of high rates of natural increase, migration has increased as a result of recession, drought, displacement from farms, and the effects of war and un- and under-employment in the northern bantustans. Overall conditions and human densities in Katutura have not improved during the 1980s, and a significant inflow of returning exiles in 1989 has further increased pressure on available shelter. However, rapid new construction and greater purchasing power of the expanding coloured middle-class have reduced pressure somewhat in Khomasdal.

The state's urban shelter policy for Africans has undergone a metamorphosis, commencing with the granting of freehold rights and efforts to sell the existing municipal housing stock to tenants on a sliding scale of prices. Residents were initially distrustful and the response was very slow for several reasons including lack of affordability (Simon, 1988a), but by early 1989 some 53 per cent of the 5,000 units concerned had been sold. Since 1981, all new construction has been for sale, not rent. As in South Africa, the cost of state shelter policies, the growing ideological attachment to the elimination of subsidies, the lack of affordability to most township residents of conventional municipal houses, and political change fostered the need for alternatives. Initial experiments with core housing, influenced by emerging Urban Foundation practice in South Africa, floundered because of attitudinal and implementational problems and convinced the conservative municipal authorities that large scale contract construction was preferable (Simon, 1988a). However, in 1982 the new parastatal National Building and Investment Corporation (NBIC) took over responsibility for low-income housing from the municipality, which now only sets minimum standards, provides infrastructure and sells surveyed and serviced plots to NBIC. The NBIC has pursued a two-track strategy, on the one hand continuing with large-scale construction of conventional township houses, and on the other seeking to gain popular and political acceptance for lower standards including core housing and self-help initiatives broadly within the World Bank mould. Various such experimental schemes and designs have been implemented in Katutura and elsewhere with mixed results, but in the main the NBIC has found itself caught between conservative authorities and popular resistance. The former remained wedded to the existing unrealistically high standards and were generally hostile to what they saw as little more than forms of controlled squatting, while many township residents found such ultra-low cost schemes demeaning and symbolic of the neo-apartheid state's continuing exploitation. Despite efforts to involve the community and tailor packages to the needs and resources of individual households, NBIC is often seen as acting high handedly (see *The Namibian*, 27 April 1990).

As in South Africa and elsewhere, many residents experience problems in meeting rent payments and bond (mortgage) instalments because of low income. Default and ultimate eviction are therefore common − at least 10 per month in Katutura alone (Eixab, 1986). Such displaced people are forced to lodge elsewhere or to erect 'informal' shelters on the township fringes (Muller, 1988). Poverty is a pervasive problem, with high adult unemployment (the 1986 Catholic Church survey put the long term figure at 43 per cent [Von Garnier, 1986]), and some two-thirds of the households living permanently below the Household Subsistence Level poverty datum line.

Even those in employment generally suffer exploitation and very low wages. Average wages in the 1986 Catholic Church survey show little absolute improvement on those found during the author's own survey during 1980−81 (cf. Simon, 1984b; Von Garnier, 1986). Over the intervening period, the *official* inflation rate, especially for food and primary goods, had stood at between 10 and 18 per cent per annum! In 1986 the NBIC calculated that over half the low income residents of Katutura who are in employment could not afford even the most basic core house then available. Nationally the figure was over 60 per cent. Existing low-income housing policy has clearly failed this large component of the urban population (NBIC, 1986; Muller, 1988; Simon, 1988b). Such findings underpin NBIC's conviction that existing standards and construction costs must be lowered in order to cater for the poorest of the poor.

Accordingly, NBIC's five-year construction programme for 1990−94 envisages far greater emphasis on ultra-low cost housing and site and service schemes in Katutura, and ultra-low cost housing in Khomasdal (Table 12.1). Conventional

Table 12.1 NBIC's construction plans in Windhoek (1990−4)

	1990−91	1991−2	1992−3	1993−4	1994−5
Katutura[1]					
Conventional	143	102	78	48	39
Ultra-low	502	549	566	588	573
Site and service	289	369	468	576	705
TOTAL	934	1020	1112	1212	1317
Khomasdal					
Conventional	256	234	213	199	183
Ultra-low cost	180	268	365	464	578
Site and service	0	0	0	0	0
TOTAL	436	502	578	663	761

Note
[1] Including the new extensions of Wanaheda and Okuryangava.

low-income housing is defined as costing over R19,000; ultra-low income housing as costing between R8,600 and R19,000; and site and service schemes as costing under R8,600, all at 1990 prices (NBIC, 1989). This strategy was drawn up before independence, so may be changed as the new government's policies are formulated. Against the background of poverty, need, dissatisfaction with state housing policy, and stimulated by the International Year of Shelter for the Homeless, the Catholic Church and NGOs such as the Saamstaan Housing Co-operative have also become active in the provision of low-cost shelter. Thus far the scale of their operations is small, although Saamstaan now has some 300 members.

In 1980, the municipality began locating some of the new housing and institutions in the former buffer strips around Katutura and Khomasdal to maximize the efficiency of land utilization and to remove these symbols of apartheid. The extent of this, although constrained by topography, can be gauged by comparison of Figures 12.1 and 12.2. However, construction by the central authorities of the Western Bypass highway in the late 1970s ensured that physical contiguity with former white suburbs could never be achieved (Figure 12.2).

Plans to close the contract workers' hostel, arguably one of the most potent symbols of urban apartheid, and rehouse its occupants date back to the first 'Katutura Emergency Housing Scheme' in 1980–81. Notwithstanding the end of contract labour and abolition of the pass laws, the hostel continued to house large numbers of men in poor conditions and under an oppressively policed regime. However, nothing changed until, in 1987, the South African appointed government instituted a crash programme to build alternative shelter and to demolish the hostel. To maximize the political capital, the cabinet chairman presided over its symbolic destruction with explosives on 9 October 1987, and the bricks were given free to Katutura residents (Simon, 1988a, 1988b). However, the ultra-low cost structures to which the occupants were moved have continued to be a source of friction with NBIC (*The Namibian*, 27 April 1990).

Infrastructure, social services and amenities in the townships remain paltry (cf. Pendleton, 1974; Von Garnier, 1986; Simon, 1988a), while the incidence of violence and theft are increasing. Open piles of rotting refuse, pools of stagnant water in summer, inadequately maintained gravel roads and negligible traffic control are also serious problems. Car ownership, though rising, is still low, but public transport is very inadequate and there is no city-wide system. A municipal bus service geared to the major commuter flows of domestic, commercial and industrial workers, links Katutura with the central city and main suburbs. Service levels and route diversity do not cater for many people's needs. A private company provides a similar service from Khomasdal. In recent years, the number of registered and pirate taxis has increased markedly, and the trade association, NABTA, is an affiliate of SABTA in South Africa. All taxis are now allowed to operate throughout the city, previous racial restrictions having been repealed through an amendment to the Road Transportation Act in 1988.

Conclusions, future prospects and implications for South Africa

By independence, the apartheid city model had become blurred to a significant degree in Windhoek. All legal restrictions on where people may live and mix socially had long been removed, and the resultant 'greying' became generally uncontroversial. Initial incidents of hostility and occasional violence subsided. One possible reason for this is that the proportion of black people living in former white areas remained small until 1989, exceeding 25 per cent in only a couple of areas. Furthermore, a significant number of people classified as coloured who initially moved out of Khomasdal because of the chronic accommodation shortage, expressed no particular desire to live 'among the whites', and subsequently returned to Khomasdal when they found housing or obtained a bond to cover construction of new and larger homes (Simon, 1986). More generally, though, operation of a capitalist land and property market has replaced legislative fiat as the regulator of residential integration. Hence the class character of individual areas has largely been preserved and property prices have generally been raised rather than lowered. There has certainly been little evidence to support notions of white flight as black people move in; on the contrary, white movement commonly preceded infiltering.

The availability of higher quality housing has to some extent been dependent on the rate of white emigration, which fluctuated during the 1980s with political uncertainty and perceptions of the prospects for, and consequences of, independence. Prices and levels of new construction varied accordingly. Many of the most reactionary elements and some commercial and professional people had left during the years preceding 1988, when the independence accords were signed. A certain number left immediately thereafter, but the flood anticipated in some circles did not materialize, not least because of SWAPO's conciliatory tone and because the situation in South Africa seemed even less predictable. Conversely, the arrival of a significant number of the 42,000 returning exiles and the influx of UNTAG personnel and foreign diplomats during 1989 and early 1990 created unprecedented pressure on accommodation of all types. By late 1989, the price of high quality housing had reached a level comparable to parts of the southern suburbs of Cape Town, for example. Although a certain amount of property was available up to *c.* R150,000 as owners emigrated or sought to hedge their bets and capitalize on market conditions, prices in the luxury market were grossly inflated. Major embassies paid up to R4–5 million in extreme cases. These trends will have increased the level of residential integration substantially, especially in suburbs like Luxury Hill, Ludwigsdorf, Klein Windhoek, Eros, Eros Park and Olympia, but will have severely reduced the ability of existing Windhoek residents to obtain higher quality accommodation outside the townships.

In broader terms, though, many apartheid structures remained largely intact. The single most important law entrenching them was the notorious Representative Authorities Proclamation, AG8 of 1980, which created ethnically based second tier authorities with powers *inter alia* over health, primary and secondary education, social welfare and pensions, and, in the former bantustans, also shelter provision. Financial allocations and the availability and quality of facilities and services thus varied just as widely as under old style apartheid. Most state hospitals and schools remained segregated, and despite longstanding overcapacity, the ultra-conservative Administration for Whites only bowed to the inevitable after SWAPO's election victory and began admitting limited numbers of black pupils to schools under its control for the 1990 academic year. There is a severe shortage of places in Windhoek's township schools, especially in Katutura.

Despite the repeal of discriminatory legislation, shelter policy and administration in the three former segregated areas remain distinct in practice. NBIC was instrumental in formulating plans for a new uniform national housing policy to replace the apartheid structures but the final document, submitted to the South African appointed government in August 1988 (Cabinet Committee on Housing, 1988), was not formally approved before the latter's dissolution ahead of implementation of UNSCR Resolution 435 in early 1989.

The Windhoek Municipality's plans to create separate neo-apartheid municipalities in Katutura and Khomasdal were frustrated by central government in 1983 (Simon, 1985), but no alternative local authority structure was forthcoming. The Windhoek City Council remains an all white preserve, and a change in the law would still be required to enable black residents of former white suburbs, let alone residents of Katutura and Khomasdal, to vote in local elections. Their only form of 'representation' is toothless and, especially in Katutura, unrepresentative township advisory bodies dating from the old apartheid era.

Immediate priorities for the new government of independent Namibia are inevitably focused on removing the remaining bastions of segregation and welding single, integrated services in areas such as government administration, education and health. Ministers are familiarizing themselves with conditions, and are stressing the need for reconciliation and deliberate change in keeping with personnel and resource capabilities. Reckless gestures are being strenuously avoided in the first few months. Abolition of AG8 was effected by the independence constitution. The reorganization of local government is receiving high priority, while commitments to the provision of adequate and affordable shelter for all have also been given.

If the experience of other newly independent states is repeated, symbolic urban changes such as the renaming of places and streets, and the replacement of colonial monuments and statuary, will occur in the near future. These are relatively cheap and easy to accomplish, and help to assert new identities and legitimize the new order. Effecting major changes to the concrete and bricks of the existing urban fabric will be far more difficult, and little in this direction has been achieved elsewhere. Some of the urgently required upgrading of infrastructure and facilities in Katutura and Khomasdal can be expected, although substantial departures from past policy will be necessary to achieve a real redistribution of resources and meet the challenge of continued urban growth in the capital. Existing conditions are unsustainable and irregular or 'informal' shelter is almost certain to become more widespread.

Contrasts and comparisons with the South African situation have been made throughout this chapter. Although legislative desegregation began much earlier in Namibia and was probably administratively easier to accomplish, low income shelter policy and reorganization of local government are now lagging behind South Africa. The South African state seems to have taken heart from the experience of desegregation in Windhoek, but conversely singally failed to learn some of the lessons from the failure of neo-apartheid administrative systems in education and health when it reorganized these in the early 1980s. Only now are these costly errors being reversed in both countries. It is significant that Gerrit Viljoen, the cabinet minister widely regarded as the architect of the dramatic changes being implemented by President F. W. de Klerk, cut his political teeth as Administrator General in Namibia during the early 1980s.

There are a number of significant differences too: the level of African urbanization in Namibia is rather lower than in South Africa, while no bantustan was situated sufficiently close to Windhoek to permit 'frontier commuting' as has occurred in the PWV, Natal and Eastern Cape metropolitan regions (see Lemon, 1982). More rapid rates of rural–urban migration by Africans may thus have occurred once restrictions were relaxed in the late 1970s, especially in view of the war in the North. However, the existence of several regional centres between the northern bantustans and Windhoek, coupled with the capital's very limited industrial and, indeed, overall employment base compared with the South African cities discussed in this book, has probably dampened urban growth somewhat. Nevertheless, the last few years have witnessed growing urban unemployment, a problem likely to persist or even worsen in the foreseeable future as a result of heightened popular expectations of the new state and the eventual concentration of a significant proportion of recent returnees in Windhoek. While the civil service payroll has already grown in the first few months after independence, anticipated new foreign and local investment is unlikely to provide employment on a scale commensurate with need. Circulatory migration (seasonal mobility) is also likely to remain significant for at least some members of many African households.

Although South Africa is a far larger-scale society, and the magnitude and complexity of urban change commensurately greater, experience in Windhoek (and Namibia generally) is pertinent to the future direction of South African cities. First, while apartheid legislation may have been essential to the forced urban segregation, wholesale integration will not follow repeal of such laws. Existing 'grey' areas may become greyer and some new ones arise, but in the main, underlying structural and economic forces, such as the capitalist land and property market and a highly unequal income distribution will inhibit the process. These relations of production and reproduction will prove far more resistant to change. The emergence of a genuine and democratic post-apartheid order will require fundamental changes of which the repeal of discriminatory legislation, however laudable, can be only an initial reformist component.

Secondly, for reasons discussed above, integration will generally occur on a class compatible basis, with the middle and incipient middle classes probably standing to benefit most. Thirdly, and following partly from the previous point, although it may initially come as something of a shock, the discovery that people of different pigmentation and/or culture have much in common will tend to dissipate many early apprehensions and forms of hostility to integration. A problematic hard core of reactionaries may well remain, but most people adapt to change when it actually occurs. One of the most reprehensible aspects of apartheid has been the ignorance, distorted racial stereotyping, fear and conflictual ideology deliberately inculcated in the cause of legitimizing the system.

13
HARARE: A WINDOW ON THE FUTURE FOR THE SOUTH AFRICAN CITY?
Neil Dewar

Zimbabwe gained independence from white settler colonial rule in April 1980. The black government that came to power after a bitter and protracted 'bush war' was explicitly Marxist in its political ideology. Indeed fears and uncertainty concerning the nature of change under this new political philosophy had fuelled white resistance to black hegemony. It is axiomatic that such change would have implications for development at both national and local scales including employment opportunities, ownership of property, the quality and availability of health services, housing, transportation, access to recreation facilities and the provision and maintenance of public services. A decade later the same political party continues to monopolize power (indeed is expressly intent upon the creation of a one-party state), and to endorse, albeit in more muted, less extravagant rhetoric, its original socialist political philosophy.

Ever since independence, events in Zimbabwe have elicited close, usually critical scrutiny from its southern neighbour, South Africa. In the current context of political flux and uncertainty in South Africa, there is widespread conviction that Zimbabwe might somehow provide a window on the future and it is in this light that significant dimensions of the developmental experience of the city of Harare (formerly Salisbury) are considered.

As Davies argues in Chapter 5, in order to adequately understand urban form and functioning under any specific mode of production, it is particularly important to apprehend the changing nature of the social formation. For present purposes, it is important to recognize that the settler-colonial social formation was reflected in the physical form and functioning of the city. The city was, like the *pays* of Vidal de la Blache, a medal struck in the likeness of its people. In South Africa white domination is still entrenched (moves toward negotiation notwithstanding) and, in terms of Davies' thesis, the dominant group is merely confronting a new crisis threshold. Conversely, at the Lancaster

House constitutional conference in London in 1990 the final 'threshold' in Rhodesia was breached by the introduction of a universal franchise.

No attempt is made here to present a morphological analysis of Salisbury as a prototype of the settler colonial city. Useful descriptions of the early development of the city may be found in Tanser (1965 and 1974) while Dewar (1988), Kay and Smout (1977) and Christopher (1970) provide more theoretical analyses of its evolution. The central focus of this chapter is the nature of change experienced in post-independence Harare. Inevitably, however, discussion of any change requires recognition of the circumstances obtaining before the event. It is therefore necessary briefly to place the city in the context of its particular social formation and to point to its more salient morphological features.

Salient features of the Rhodesian social formation

In order to adequately comprehend the nature of the evolving settler colonial political economy of Rhodesia, it is important to recognize that the pioneer column that penetrated the territory comprised citizens of the Cape Province of South Africa, not people directly from England. Their ethos was strongly influenced by earlier experiences of the Dutch and British settlers in South Africa in dealing with the 'kaffirs' and in mastering their physical environment, as well as by a strong conviction about the superiority of British technology, laws, religion and morals. They brought to Rhodesia 'the colonial institutions, attitudes and habits, and especially the racist stereotypes of the Africans which, in turn, explains some of the injustices and irrationalities that have characterized the administration of Southern Rhodesia' (Chanaiwa, 1981, p. v.)

From the outset, white commercial and political interests were protected by institutional, legal and coercive practices. While blacks were deliberately drawn into the cash economy through a variety of strategies (Arrighi, 1970), other strategies, notably the control of education, job reservation, regulation of trading practices and the limitation on technical skills training, served to entrench the economic supremacy of the white group (Phimister, 1975; Utete, 1979). Over time, class ascription by the dominant group increasingly became a function of race, while race differentiation became increasingly subject to social and legal prescription. Values had become associated with motives; ideology became entrenched in institutions and legal structures. Dominant group bureaucrats became gatekeepers and managers pragmatically practising the politics of exclusion and legitimizing their actions on the grounds of 'civilizing' or 'developing' backward peoples. Differences and discontinuities were emphasized, forming 'the social mechanism by which inequalities are maintained and incipient conflict regulated' (O'Callaghan, 1977b, p. 12).

Domination, separation, segregation and political and economic inequality on a national scale were continuing features of settler-colonial society. These processes were embedded in numerous laws and were maintained both by an extensive white bureaucracy and by social conventions. The extension and intensification of institutionalized discrimination in Rhodesia is well documented

(Plewman, 1958; Palley, 1966; Quenet, 1967; Austin, 1975). A pervading theme in these and other sources is differential access to resources at every level of social, economic and political life.

In the political sphere, blacks were largely unenfranchised and ill-represented on critical decision-making bodies. In employment there existed marked pension and wage disparities; for a long time blacks were precluded from forming unions and those that did exist were segregated; they were also denied the opportunity to acquire skills through apprenticeship. Black educational provision, housing and living conditions were demonstrably inferior to those experienced by whites; blacks possessed neither residence rights nor choice of accommodation. Physical and social mobility was severely constrained. Recreation facilities were more limited and invariably of lower quality than those available to the white group.

Salisbury: a reflection of Rhodesian culture

In general Salisbury manifested the political economy of capitalist settler colonialism and its structure, form, and functional organization conformed with the general model of the colonial city (King, 1976; Simon, 1984a). In addition, however, it reflected the circumstances of its own history and geography. Race discrimination became a primary determinant of social, economic and political structures. Born of a need to entrench white settler dominance over the subordinate indigenous population, discrimination subsequently served to reaffirm and to reinforce white attitudes and values which were the bases of behavioural norms and practices. A uniquely Rhodesian culture was reflected in spatial structures and physical forms which, once established, became geographical imperatives determining many facets of city life. Racial separation, segregation and inequality also became pervasive features of the city's geographical environment. The physical segregation manifest in geographical space created and emphasized cultural differences which further affirmed and justified functional discrimination. The spatial plan of Salisbury determined, entrenched and perpetuated social attitudes, relations, and activity patterns.

From its inception Salisbury was a planned town, its design and layout revealing late nineteenth century planning principles drawn most immediately from South Africa but, more generally, from Britain. The influence of the Garden Cities movement was particularly evident; land use separation, single detached houses on large plots (seldom zoned less than 1,000 square metres and, for large areas, a minimum of 4,000 square metres), wide tree-lined streets, extensive parks (the 'country-in-the city' ethic) and the locational and functional dominance of squares and public buildings which symbolized and constantly reaffirmed the primary settler purpose of politico-administrative dominance (Western, 1985). By the late 1970s with a population of approximately 569,000, the city contained over half the country's production enterprises located in seven major zoned industrial sites while the white population was well served by retail and service outlets in 66 conveniently located, planned shopping centres excluding the central business district (Kay and Smout, 1977).

The original design of the city took no cognizance of an ubiquitous black presence. It was conceived and developed as a white urban settlement. Nevertheless, many blacks who had gravitated to the settlement seeking work had congregated on the outskirts of the settlement. Other labourers occupied *kias*, small outbuildings on the properties of white residents. In 1907 the municipality assumed formal control of the spontaneous black settlement. It became the first black 'location' (township). Consistent with South African and British colonial traditions it was to remain entirely segregated from the white residential areas (in this instance the railway line and Makabusi river formed physical buffer zones). Subsequent provision of land was made on an *ad hoc* basis for African needs. Additional townships were planned to extend as wedges to the south and west, and later to the east, adjacent to the railway line and in proximity to the industrial land, but always geographically separate from white residential areas (Figure 13.1). Between 1950 and 1978, six major townships were planned within the greater Salisbury area.

Each township was planned according to the planning precepts (notably that of the inward-oriented neighbourhood unit) and technology obtaining at the time and in conformity with white planners' perceptions of black needs. Housing quality was an index of the ability of low-income families to pay for what they received through the payment of rents and service charges. Cost was an overriding consideration as, by law, each township had to be entirely self-financing.

Commercial and industrial activity, both formal and informal, were dynamic elements in the functional organization and physical fabric of the black townships. Formal shopping centres were planned as functional hierarchies and distributed throughout the townships but these generally provided only very low-order functions. Centres differed markedly in the black and white areas in terms of scale, facilities, design, maintenance, aesthetics, and range and type of goods offered. Limited provision was made for open-air trading in market stalls and by licensed vendors operating from prescribed sites. Hawkers and vendors were subject to stringent police control but many operated illegally. Petty commodity producers and service traders could be found in approved Council premises. Others operated on a more temporary basis on vacant land and in back yards.

The privileged socio-political and economic position of whites was reflected in the type of houses and services available to them in the proclaimed white areas of the city. High average income and political power were factors that considerably enhanced their quality of life. Conversely the black areas were products of relative poverty being unable to attain the necessary thresholds to support an extensive, high-quality provision of public goods and services. Moreover blacks had no representation on the City Council. Instead they were regarded as comprising an entirely separate system with its own administration and financing. Resident representation was limited to Boards (part nominated, part elected) which until 1978 possessed only advisory powers (Figure 13.2). Rising conflict thresholds resulted in progressively greater compromise and appeasement from a situation of complete white control by a supervisor to

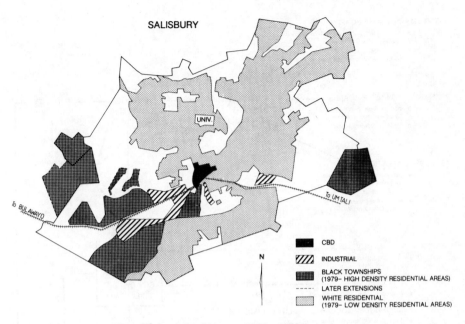

Figure 13.1   Structural outlines of Salisbury, 1978–1979

extensive representation and executive authority for blacks in proclaimed townships.

As a function of their higher incomes, whites were able to support a wide range of urban amenities and facilities. More importantly, while benefiting from the municipal supply and maintenance of such physical infrastructure as electricity, water-borne sewerage, water reticulation, street lighting, tarred roads, curbs and pavements, and storm-water drainage, they were able to substitute private sector supply for public social services such as education, health and recreation. In this way they greatly increased the range and quality of goods and services available to themselves.

It should be emphasized that the actions of the settler State and of the white local authority prior to independence were not entirely without merit. Many positive and innovative initiatives were undertaken by highly competent and dedicated officials in the spheres of health, housing, finance and community services amongst others. Nevertheless it is valid to assert that the white administration did not meet the explicitly articulated need of the black populace to become permanent and effective participants in urban society. Nor were they concerned to move sufficiently aggressively toward greater social equity and the narrowing of inequality.

Transformations

Such concerns became central issues in the new Zimbabwe. The City Council's attention and energies shifted overwhelmingly towards the former black town-

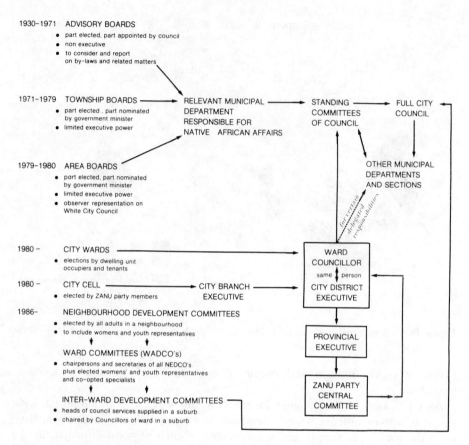

*Figure 13.2   Channels for participation by blacks in local government in Salisbury/
           Harare*

ships, now known as the high density areas. The major changes that have
occurred in Harare are largely the direct result of local authority initiatives
directed towards improving both the quality of life and the life chances of the
city's poorer residents.

A differential rates tax, whereby the tax burden upon commercial and
industrial properties increased from 29 to 39 per cent and 18 to 27 per cent
respectively, was immediately introduced. This demonstrated the resolution of
the new government to generate and to redistribute a greater proportion of
wealth from capitalist enterprises. In addition their waste removal charges were
increased by 20 per cent, water tariffs by 23 per cent and vehicle licences by 15
per cent. Also in the first financial year, health services benefited from a 25 per
cent subvention increase from rates assessed on residential properties in low
density (formerly white) areas, money previously dedicated to improvements
only in white areas. Finance was also divested from the trading accounts to
underwrite the new priorities directed towards the poorer inhabitants of the

high density areas. Capital and revenue funding on the provision of electricity, water, water-borne sewerage, health and education increased markedly in the high density areas.

Housing

Of all the public goods provided by the municipality, housing had the greatest impact on the lives and consciousness of the blacks. Previous lack of choice inherent in having housing allocated in segregated areas by the local authority, being obliged to pay rental, and not being able to own property in most townships, had symbolically reinforced the peoples' perception of their sub-servient position within the colonial political economy.

Thus housing was potentially a key element in implementing changes that would demonstrate new socialist priorities (Lancaster House protection of private property in the low density suburbs notwithstanding). The opportunity existed to abolish private housing as a commodity in all existing and future high density areas and to place this housing and land under State ownership and centralized control to be allocated according to need and principles of social justice. The result would have been socio-economic convergence (variations by occupation and privilege notwithstanding) and greater equality of distribution. No such radical route was taken, but the Council found itself unable to finance or build houses at a rate commensurate with demand (by April 1987 the Council waiting list for houses stood at 31,628 families while the previous year only 1,361 families had been allocated houses [City of Harare, 1987]), so it adopted a policy of mobilizing the private sector in the housing delivery process. The principles of freehold tenure and economic returns to investment were embraced. Strategies included transfer to sitting tenants at favourable rates and aided self-help on land purchased through council loans. By 1985, some 80 per cent of former rental housing had been sold to individual owners. Private sector involvement is reflected in the encouragement of 'tied housing' (provided by employers for employees), which was once an anathema to the black political leadership.

The debate on the form future low-cost housing should take was never centred on the issues of tenure but rather on standards of private housing. Plot sizes, housing qualities, and construction techniques were important elements of the debate; site and service, core housing, aided self-help and the use of building brigades were strategies employed at different times (Teedon and Drakakis−Smith, 1986; Davies and Dewar, 1989). Meanwhile, expenditure on the repair and maintenance of low-cost housing has increased. While higher standards of dwellings and services in the still exclusively black high density areas have marginally shifted the housing system as a whole towards convergence, structural inequality remains strongly characteristic. In the location of new low-income, high-density housing schemes, traditional spatial approaches have been followed and the high- and low- density components of the housing system remain disaggregated. Thus most people will continue to be barred from large parts of the city on economic if not racial grounds. Though now blurred in terms of racial divisions, the strong class cleavages of the past are tending to intensify spatially.

As a condition of the Lancaster House agreement, private housing in the former white suburbs was protected from expropriation for 10 years. Following the repeal of the Land Tenure Act by the shortlived government of 'Zimbabwe-Rhodesia' in February 1979, a movement of more affluent blacks from the high-density areas to hitherto exclusively white, low-density suburbs began. After independence black political leaders, civil servants and many diplomatic missions from African countries entered the housing market as owners and renters (Figure 13.3) (Davies, 1988). Areas receiving relatively large numbers of blacks in these early years included the relatively low-income southern white suburbs of Parktown, Prospect and Midlands, the mainly flatted area of the Avenues near the city centre, and Mabelreign, a 'garden city' suburb with many smaller properties in western Harare (Harvey, 1987). The movement of these blacks provided a demonstration effect that further mitigated against eventual implementation of a socialist housing delivery system. By 1985 the ownership of approximately 22 per cent of dwellings in the low density areas had been transferred to African owners (Davies and Dewar, 1989).

Property values in the high density areas were depressed at the time of independence, in part because of white emigration. In a case study of Mabelreign, Cummings (1991) finds that low prices allowed whites to move more easily to higher-income suburbs for a while, but that large-scale movement of blacks appears to have been delayed by the time required for knowledge of the changing housing market and its mechanisms to diffuse among potential black buyers. The fear of a hostile reception by white residents may also have played a part. Within Mabelreign, as in the city generally, blacks moved in the greatest numbers to those districts with the lowest average property values. Nevertheless the black component of Mabelreign increased from less than 10 per cent in 1980 to c.55 per cent in 1984.

As the post-independence exodus of whites slowed down, and the demand for housing from a rapidly expanding middle-income black sector increased, house prices began to rise in the mid-1980s. By now few whites were moving to other suburbs or leaving Zimbabwe, so black in-movement slowed down, and the black component of Mabelreign had increased only to c.63 per cent by 1986. By this time Cummings points to an emerging shortage of middle- and even higher-income housing, and the development of new middle-income housing schemes to meet the demand.

As in South Africa, the potential diffusion of blacks into the white residential suburbs with its implications for mixed schooling and the utilization of local public facilities especially schools, clinics, libraries and swimming baths had traditionally been a source of concern for whites. So too was the fear of lowered values and the somewhat more ill-defined apprehension of a lowering of 'standards' (overcrowding, ill-kept gardens, lack of maintenance, noise and possible 'unsavoury' activities). Despite such fears, the movement of blacks into the low density areas has not provoked overt reaction from white neighbours, although very little social intercourse occurs between the two races

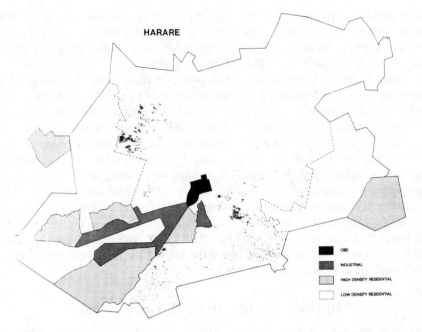

HARARE

CBD

INDUSTRIAL

HIGH DENSITY RESIDENTIAL

LOW DENSITY RESIDENTIAL

Figure 13.3   Transfer of dwellings from whites to blacks in Harare, 1979–1985

(Ribnick, 1987). Suburban shopping centres and public facilities are used extensively and unselfconsciously by blacks as the need arises. Very few of the household members surveyed had joined local sports clubs, yet few retained contact with residents and organizations in the high density areas, most seeming to make a new life for themselves in what was white Salisbury. The ability of wealthier blacks to move in this way has intensified the spatial expression of class cleavages. Property values in the low density areas did not decline subsequent to black infiltration. Notwithstanding considerable white emigration on political grounds after independence, pent-up demand from a large, relatively affluent black middle class and newly created elite, in addition to the staffs of numerous national embassies, consulates, aid and other organizations arriving at the onset of independence, mitigated against the creation of a buyer's market and falling prices.

Squatter camps had long been present in and around Salisbury. Immediately before independence, squatter camps became reception areas for increasing numbers of migrants and refugees from the rural areas as the liberation war intensified. Until then their proliferation had, however, been contained by the municipal authorities as a result of constant monitoring and intermittent clearance programmes, many of which were considered oppressive and elicited vehement opposition from diverse organizations in opposition to white minority rule. Members of the Harare City Council nevertheless also considered squatting to be undesirable and potentially a major problem. Comments and opinions from black councillors were often forthright: '... people who have jobs can come:

those who have nothing to do must do it (sic) in the communal lands'; 'Many are vagrants. Half to three quarters are not there because they have no accommodation. Some get homes, rent them, and squat. They are crooks. They brew illegal beer ...' (Dewar, 1988, p. 404).

Council policy is that no one may squat. Squatters are screened by the Department of Housing and Community Services to ascertain whether or not their need is 'genuine'. Attempts are made to provide accommodation for those with productive employment while those without are treated differently according to their circumstances. Immigrants and refugees become wards of the Department of Foreign Affairs prior to repatriation. The unemployed become the responsibility of the Department of Lands and Resettlement and nationals are returned to their place of origin in the rural areas or to rural land resettlement schemes. Councillors strongly endorsed government policy of developing rural growth points and of improving conditions in the rural areas in the belief that this would diminish the relative attractiveness of the capital city. Their primary concern was to maintain standards in accommodation and residential environments.

Improving the high density areas

The City Council's efforts in this sphere are wide-ranging. In addition to housing and physical infrastructure they embrace education, health, libraries, utilities, transport, sport and recreation, drinking facilities, the promotion of small-scale enterprise and co-operatives, and crime prevention.

The Council has assumed responsibility for both pre-primary and primary school education. This involves adherence to the government's integrated National Early Childhood Development Curriculum which targets six dimensions of child development: physical, cognitive, creative, social, emotional, moral and health development. 'Hotseating' (the use of classrooms for two successive groups each day), an infant feeding scheme and the active involvement of parent associations are all features of the programme. Provision of a united, city-wide library service is also under way.

The provision of both preventative and curative health care has been considerably increased. A significant innovation has been the creation of seven polyclinics with centralized multi-unit facilities. Many new clinics and primary health care facilities have also been opened in the high density areas. Free health care for the indigent and those earning less that Z$150 per month (c.£30) has been instituted. The biggest constraint upon the quantity and quality of health care is, inevitably, lack of funds and trained personnel.

An electrification programme in the high density areas is almost complete. The Council is also vigorously pursuing its commitment to provide all houses with running water, power and light. Amenity-related expenditures which have been increased in real terms since independence include the upkeep of recreation grounds, grass control, environmental health and the upkeep of roads. Significantly, such expenditures have also grown (less rapidly) in former white suburbs: this reflects both a concern for the maintenance of 'standards' by black councillors, and demands for a continuing high quality provision of facilities and amenities from the thousands of blacks who have moved into these areas, and who form a new pressure group.

In 1987 the City Council acquired the National sports centre and centralized the administration of sport. Its programme includes the training of coaches and sports club management and the conducting of clinics, the aim being to broaden the range of sports available to the people. By 1985 the Council had supervised the formation of 95 junior clubs in the high density areas, providing facilities for many sports which had previously been the preserve of white participants.

Drinking facilities are being steadily upgraded to make them more socially acceptable and comparable with those in the city centre. Many older public bars are being converted to neighbourhood taverns. The Council also intends to provide entertainment facilities in its liquor outlets, to diversify its operations into the low density areas, and even to move into the night-club business.

As a contribution to law and order, the Council has enlisted ex-combatants of the 'bush war', including a women's detachment, into a municipal police force tasked particularly with patrolling council liquor outlets, guarding municipal properties, maintaining order at bus termini, patrolling stream banks for illegal cultivators and assisting the national police (although with limited powers).

Although the State has acquired ownership of, or part share interest in a wide variety of formerly privately owned commercial and industrial enterprises (Zinyama, 1989) there is no evidence of prohibition or restriction upon private, profit-motivated enterprise as might have been expected under a socialist regime. Indeed the Harare City Council is acting as a catalyst for the promotion of small-scale marketing and petty commodity production. This is accomplished in a number of ways: by developing sites and facilities for formal shopping outlets and for 'people's' markets; by removing earlier colonial-era restrictions on hawkers and vendors and by the extensive licensing of 'emergency taxis'.

The Council's Department of Housing and Community Services endeavours to promote the formation of viable profit-motivated co-operatives in service industries, commodity retailing, and agriculture on unused municipal land and to educate the public about co-operative activity. It has also started its own silk farm while selling beef, maize, and vegetables produced on five other Council farms. Income derived from these operations is channelled to welfare-related expenditures. The Council undertakes the training of womens' organizations to improve craft skills with a view to marketing. Stemming from these initiatives is the development of public responsibility and self-reliance within the community. The creation of Council building brigades (their relative lack of success notwithstanding) and youth development through the organization of craft workshops are further examples of this.

The considerable reallocation and geographical redistribution of public authority finance evident in the published accounts masks the tenor of the Council's resolve to redeem past inequalities. Underspendings on the revised estimates regularly exceed 50 per cent. This occurs for a variety of reasons, mainly shortages of foreign exchange for purchasing plant and equipment, skilled manpower shortages, and the deferment of the land purchases particularly for housing schemes. Generation of finance to attain well defined goals is a constant problem.

Government attempts to mobilize and raise the consciousness of the 'povo' (the masses) are pervasive in the mass media. In Salisbury, during the war years, a system of cells and branches was put in place in order to channel directives and to incubate party principles and goals. Later, while this primary function continued, the system was used as an instrument in citizen decision-making, a conduit for popular expression. However, because these channels of communication were exclusionary insofar as they were confined to party members and because the system tended to bypass the Council and to diminish its effective authority, the institution of Ward Committees with the City Councillor as Chairman was established after 1980 (Figure 13.2). These are apolitical and possess limited executive powers including the right to generate finance for local community projects. This capability further points to the considerable energy and resources being directed toward improving the material and spiritual/aesthetic circumstances of black residents in the high density areas and to instilling a spirit of responsibility and self-reliance.

Conclusion

The ultimate verdict on change in Harare after a decade of majority rule is that the physical structure, form and daily functioning of the city has undergone minimal transformation. The communications networks, land-use patterns and activity patterns established under settler colonial rule largely persist. The burgeoning black lower income residential areas form a physical extension of the earlier segregated black townships. No attempts have been made at achieving a more desirable socio-economic mix of the population. Despite the emphasis on redistribution and possibly because of the considerable political leverage of large numbers of new middle class black residents in the low density suburbs, provision of public goods and services has not fallen below pre-independence levels.

However, a determined redistribution of wealth away from the former white residential, industrial and commercial areas towards the more disadvantaged high density areas is being effected through the reallocation of municipal funds and resources. Nevertheless, this allocation constitutes a change of emphasis within a relatively unchanging urban system.

While blacks constitute some 95 per cent of the work force in the public sector (Davis and Dewar, 1989), whites still continue to dominate the wealth-generating commercial and industrial sectors through ownership and occupation of managerial and high-skill positions. Entry to managerial occupations for blacks in the private sector remains constrained. The National Manpower Survey of 1984 observed that, 'it is fair to conclude that a good deal of black promotions .... since 1980 have been cosmetic window dressing' (Zimbabwe, 1984, p. 2). At the social level little interaction occurs between the ethnic groups. Shortages of foreign exchange and consumer goods aside, the most significant change in lifestyle for whites is a greater reliance on the private provision of education, health care and recreation.

Certainly no radical structural reorganization of the urban space economy, of the sort effected in the Eastern bloc countries under socialist governments after World War II, has occurred. Why then the disjuncture between the intent of implementing a socialist model of development and the evident absence of concomitant structural and spatial transformation? The answer lies in a complex mix of factors, but five major reasons which apply as much to the South African scenario as to Harare, may be advanced.

In the first instance, the position of the city in its regional and national context must be recognized. To the extent that the forces operating at the national level to inhibit reform initiatives are translated to the local level, freedom of action is compromised. In Zimbabwe, the government fixed interest rates, determined foreign currency allocations, directed major capital investments, negotiated foreign loan-and-aid packages and prescribed the nature of local authority loan applications. Importantly, it also pursued a conscious policy of rural development at the expense of the urban areas and, by statute, greatly circumscribed Harare's freedom of action in raising and spending money. In consequence, almost all the City Council funds are non-discretionary.

Secondly, there is the factor of black perceptions and expectations. Prior to colonial penetration, blacks had not experienced Western urban industrial society. The development of towns and cities was a necessary concomitant of colonial political and economic domination and was a prerequisite for the functioning of capitalist economies. As a result, whites created a strong demonstration effect, by virtue of their political dominance and high material standard of living. Given that peoples' perceptions and aspirations are, to a considerable degree, a function of their own experience, the institutions and practices that had ensured and characterized an efficient space-economy provided the only practical model to the masses of the people of what was possible and, by extension, desirable.

After independence the political leadership was conscious of the need to redress deep-seated grievances that had fuelled the war effort: racial discrimination, segregation, inequalities in education, health and job opportunities, land alienation and political powerlessness. While many of the political and academic opinion-makers perceived capitalist imperialism and its manifestation in the political economy to have been the structural cause of black impoverishment and oppression, the majority perceived them in a more immediate and pragmatic manner. Practically it was easier to address the immediate problem of redistribution and development than to transform the structural framework within which these objectives might later be achieved. The financial, legal and organizational instruments used to implement public policy were efficient and, above all, in place. This placed a considerable inertia upon change.

Allied to this there existed a powerful and deeply engrained municipal corporate culture. The research, assessments, planning and recommendations of specialist departments are powerful influences upon councillors' decision-

... g. Given its antecedents, the municipal bureaucracy retained a very Western conception of how the urban system should be managed. Acting as 'gatekeepers and managers' they constituted powerful forces of conservatism and inertia. Thus it was possible to have an ideological, theoretical and political realignment within the city's power base but not an effective functional realignment.

A third crucial factor inhibiting change was the degree to which future expenditure was already committed by what had been spent in the past. The colonial city represented a vast stock of existing capital expenditure. Given limited budgets and the rate and scale of urban growth, it was not feasible to write off existing investment. Also, to a great degree, this dictated what had to be spent on repairs and maintenance in the future. Moreover, the logic of technological efficiency frequently meant that one phase of development determined the nature of the next phase. The legacy of the built form and the functional organization of the colonial city thus constituted a considerable impediment to change.

Fourthly, the Council faced considerable difficulty in breaking the behaviour patterns of people historically engaged in survivalist strategies. The urban poor, living on the economic margins and with limited physical mobility, had little option but to respond to existing patterns of spatial behaviour. The problems of the Third World city are immediate and pressing. They do not simply disappear with a change of government — they remain constant beyond ideological conviction. As a result the municipality had little option but to respond incrementally to ameliorate the immediate circumstances of the poor.

Finally, in the absence of autocratic and rigidly applied coercive control, there existed a considerable 'downside' risk associated with change. For government nationalizing the productive processes, determining patterns of consumption, and bringing land and property under state control could have incurred considerable social and political costs including jobs and functional and economic inefficiency. The nature of changes necessary to convert to the structure, form and functioning of a socialist city might have incurred costs that could not be met. Revolutionary theory could not accommodate the economic, social and political realities of the City of Harare in developing 'Third World' Zimbabwe.

Thus, at the intra-urban level, socialist theory was never extended to the articulation of planning goals and objectives for structural and functional reorganization towards realization of a socialist city. While redistribution of public funds towards a greater provision of welfare-related public goods and services in the former black townships has occurred, large-scale economic and spatial restructuring has not taken place.

14
TOWARDS THE POST-APARTHEID CITY
Anthony Lemon

The immediate future

There is a curious if inevitable unreality about the actions of a 'late apartheid' government. Whilst moving ever closer to negotiations intended to lead to a new constitution and some form of majority rule, until 1991 the government presided over an apartheid system which was still intact in many of the areas most important in the daily lives of urban blacks, and continues with piecemeal changes, dismantling some measures but still resting content with incremental reform in more sensitive areas. Guarded and ambivalent responses to changing realities were presented as a long-term vision of future options. Yet while the present government remains in power, and the implementation of a negotiated constitution remains some way off, its actions will continue to affect urban futures.

Whereas the 1988 Free Settlement Areas Act may be viewed as an attempt at damage limitation, the year 1990 saw the first tentative steps towards the dismantling of the apartheid city. The first actual repeal of an apartheid measure during the de Klerk presidency was the opening of hospitals to all races in May 1990, something sought by the Mass Democratic Movement (MDM) in its defiance campaign in the run-up to the September 1989 election. Legislation for a united health service was expected to follow soon after.

Another MDM target, the Reservation of Separate Amenities Act, was finally repealed in June 1990 (with effect from October). Transvaal councils controlled by the Conservative Party since the 1988 local elections had used the Act to reverse desegregation of facilities, which had caused their towns to suffer from effective black consumer boycotts and loss of investment (Lemon, 1990b). Their actions had embarrassed the Botha government, which was clearly afraid to repeal the Act for fear of losing further right-wing white support, but its retention would have contradicted de Klerk's far bolder stance. It remains to be seen whether these Conservative councils seek to circumvent the legislation by privatizing facilities or by other means.

In the sensitive area of education, the government has been much more cautious, responding essentially to the specific pressures arising in racially mixed residential areas, where black children were unable to attend white schools threatened with closure because of declining rolls. The official response has been twofold: first, to offer more generous subsidies to integrated private schools in such areas (including state schools which re-open as private schools, as the Johannesburg Girls High School has done); and secondly to allow white state schools to accept black pupils from 1991, but only with overwhelming white parental support, and subject to other provisos yet to be finalized at the time of writing. Clearly these developments do little to meet black demands for a single, non-racial education system, which may well have to await the election of a post-apartheid government.

There are two major areas where change has been promised by the present government: group areas and local government. Longer term implications of fundamental change are considered below, but more immediate questions are raised here. Early in 1990 the government's position evolved to include an ambiguous endorsement of applications for the opening of entire cities as free settlement areas. This depends on amendment of the Local Government Affairs in Free Settlement Areas Act to permit the election of non-racial local governments in these areas, something which is certainly possible, but is likely to be overtaken by repeal of the Group Areas Act itself, promised for the 1991 parliamentary session. President de Klerk has spoken of replacing the Act with 'non-discriminatory legislation that is generally acceptable', and of retaining the principles of private ownership and security of land tenure except for development purposes (*Weekly Mail*, 27 April–3 May 1990). His concerns here probably rest on the prevention of overcrowding and the maintenance of the socio-economic character of suburbs. This would undoubtedly be perceived as racist, and might be difficult to ensure in practice. On the other hand, the concentration of mixed settlement in a few areas which has caused such over-crowding in Hillbrow, Johannesburg, would be less likely if all areas were legally open. The Harare experience (Chapter 13) does suggest that multiple occupation of dwellings can be avoided in the absence of any race zoning, and this may prove true at least in those South African cities facing less strong pressures than Johannesburg.

Official intentions for local government are less clear at the time of writing. Speaking to the Cape Municipal Association in March, President de Klerk said the present local government system would have to be replaced by a non-racial one, but insisted that this must offer 'minority protections' (*Weekly Mail*, 25–31 May 1990). This suggests that further reform of local government may become a testing ground for national constitutional change. Options under official consideration appear to include giving some of the powers now exercised by municipalities to 'neighbourhood committees', which would give more power to suburban areas to block black majority decisions. A local option, allowing towns and cities to negotiate their own local government system within prescribed limits, is another possibility.

It is difficult to see anything short of total repeal of group areas legislation and the introduction of non-racial local governments responsible for entire cities proving acceptable to blacks. The government must be aware that the time for imposed, top-down incremental 'reforms' has now passed. For anything short of the outright repeal of apartheid legislation, negotiation has become a fundamental prerequisite of popular support.

The government is naturally concerned at this stage to avoid doing anything which will undermine white support for negotiation leading to fundamental constitutional change. Whilst survey evidence of white reactions to desegregation is complex and often contradictory, probably the most significant point is made by the Urban Foundation (1990a, p. 18), which stresses that 'the role of government and leadership appears to be crucial to white attitude formation in South Africa, as has been found with the repeal of other racial legislation.' White attitudes seem to adjust relatively easily once changes have actually taken place. Outright repeal of the Group Areas Act and the introduction of non-racial local government by the de Klerk government would have two major advantages: it would ease the path of negotiations, and whites would have time to adjust to these changes instead of facing them at the same time as many others in the early post-apartheid years.

Beyond group areas

It is clear that the Group Areas Act can no longer be effectively enforced in the face of the demographic and property market forces which have arisen from an economy which is increasingly racially integrated. Implementation of the Act caused massive dislocation in many cities; will its repeal therefore be a prelude to a second major transformation? The effects of repeal in itself seem likely to fall well short of this, but it will prepare the ground for new urban management policies which would create fundamentally different post-apartheid cities.

The sheer ruthlessness with which neighbourhoods were demolished and cities remodelled in the 1960s and 1970s guarantees that, to quote Wills' words in Chapter 6, 'the shadow of apartheid planning will be evident in the geography of the city for years to come'. Whereas the implementation of legislation was an active and ruthless process, repeal is in itself passive: it allows the market to embark on a gradual correction of previous distortions, but it neither relocates affluent suburbs nor brings the poor in from the periphery.

It is sometimes forgotten that blacks can only move into white areas as whites move on. Such movement could be within their current suburbs; there is certainly scope for intensification through subdivision of plots or conversion of large houses to flats, without disastrous deterioration of living environments. It could also mean either movement to other housing areas, or emigration. The latter will naturally depend on the actions of a post-apartheid government and the wider social, political and economic climate, but it should be remembered that most white South Africans, especially Afrikaners, are not first or even second generation settlers like white Rhodesians, nor do they include significant numbers of expatriates such as the white South Africans working in the Namibian administration prior to independence. As Krige points out in Chapter

7, whites in cities such as Bloemfontein where the white population is predominantly Afrikaans are less likely to leave (although many would like to find acceptable employment in the major metropolitan areas). Low-income whites of all language groups, including some 600,000 Portuguese, tend to have fewer marketable skills and limited opportunities for emigration. Those more able to leave are likely to feel less threatened by the new order, although some will undoubtedly move for a variety of reasons. Whites will also be amongst those who come, some to exploit new economic opportunities, others to make good skill shortages which are a legacy of apartheid, to work with aid organizations, and perhaps to overcome teacher shortages in a unified education system, as in Zimbabwe.

The possibility remains that whites will move out of those areas which prove most popular with blacks. Some would do so anyway, as part of the normal pattern of inter-urban and intra-urban movement, leaving opportunities for prospective buyers of all races. The critical question is whether black in-movement will precipitate a white exodus, as in the United States and elsewhere, where 'tipping points' (usually 10−15 per cent of the incoming group) have long been recognized. Bond (1990), stressing the dangers of a free market housing regime, points to signs of 'blockbusting' by estate agents (who introduce black buyers in the hope of inducing white flight, which enables estate agents ot buy cheaply and sell dear) in Windmill Park (Boksburg) and elsewhere. Hart (1989) uses the case of central Johannesburg to argue that ghettoization has just begun, and that the processes accord closely to the early stages of segregation found in other countries. Given the very recent and partial removal of legal controls, he believes that the process of ethnic residential selection is in its infancy in South African cities. He therefore questions the relatively optimistic conclusions of those who have examined the ethnic mixing which has so far occurred in Harare, Windhoek and Mmabatho-Mafikeng (Chapters 11−13; Wills, Haswell and Davies, 1987; Pickard-Cambridge, 1988b) and in Johannesburg and the PWV region (Chapter 9; Pickard-Cambridge, 1988a; Schlemmer and Stack, 1989). All these studies reveal a modest, incremental pattern of racial residential change with few signs of conflict or racially motivated out-movement by whites.

Precisely because the process is at an early stage, prediction is difficult. Saff (1990) rightly points to the necessity of studying the micro-dynamics of each area: the character of the area and its housing (houses or flats, owner-occupied or rented), the class of those moving in, the ability and will to apply slums legislation, the availability of alternative accommodation (in areas such as Hillbrow there is a trapped element which is unable to move), and questions of land and affordability. Central Johannesburg is exceptional in the southern African region in the pressures it has experienced, and is unlikely to be a model for most urban areas, including Johannesburg's own suburbs such as Bertrams (Rule, 1989) and Mayfair (Fick, de Coning and Olivier, 1988).

Assessment of future trends also demands greater attention to black priorities. It is not only housing shortages in their own areas which have led blacks to risk prosecution and endure surveillance by moving into white areas; research by the Urban Foundation (1990a) has shown, not surprisingly, that major incentives are the marked locational and environmental attractions of white housing areas. Hitherto many Africans have preferred the lifestyle and social contacts of the townships (Saff, 1990), but this could change as open cities make larger-scale movement to former white areas easier. The Africans most likely to move to central city areas are the so-called black 'yuppies' who do not own their own homes, especially single men, given an acute shortage of single accommodation in the townships, whilst poorer blacks from rural areas may also seek whatever shelter they can find in central cities for lack of any alternative (ibid.).

Among more affluent black families, it currently appears that a higher percentage of coloureds and Indians can afford to move to white suburbs and wish to do so. But given the size of the African population base in the cities, a very small percentage moving to white suburbs could transform their racial composition, as in some of Harare's formerly white areas. A rapid rise in the absolute size of the African middle class in likely to follow a change of government, as a degree of Africanization occurs. The ANC's problems in finding accommodation for up to 20,000 exiles expected to return during 1991 point to future needs for an expanded housing programme to cater for black middle class families, a situation which Cummings (1991) has identified in Harare.

The problems and patterns of residential desegregation are largely the concerns of those who can afford to own or rent in the private sector. They are a decreasing proportion of urban South Africans. For the majority of Africans in particular, housing concerns will be dominated by quite different issues, to which we now turn.

Land and housing in the post-apartheid city

Measurement of South African urbanization is notoriously difficult, but if functional and density criteria are used, South Africa's population was 57 per cent urbanized in 1985 (Urban Foundation, 1990b). This was only slightly below average for countries with similar levels of economic development, which suggests that decades of state control have largely displaced rather than prevented African urbanization. By 1990 about half the African population was urbanized, with 39.2 per cent in the functional metropolitan regions of PWV, Cape Town, Durban-Pietermaritzburg, Port Elizabeth, East London, Bloemfontein-Botshabelo and the Orange Free State goldfields (Simkins, 1990). If those people living in 'dense settlements' in the bantustans (living at urban densities, with little or no access to agricultural land, but with negligible urban amenities) are included, the urbanized African population rises to at least 58 per cent (ibid.). Projections of future African urbanization must estimate the speed with which fertility is likely to drop, and the extent of (informal) African in-migration from neighbouring countries, attracted by the perceived opportunities of a more developed economy. The Urban Foundation (1990c) projects even faster growth of some 600,000 Africans per annum in metropolitan areas, leading to an African population in these areas of about 17.7 million by the year 2000 and 23.6 million in 2010. Two-thirds of this growth will result from natural increase rather than migration, which suggests a limited role at best for regional policies designed to reduce pressures on metropolitan areas.

In terms of housing needs, even present figures are daunting. Research for the Urban Foundation suggests that in 1990 more than 7 million people live in informal housing, including 2.5 million in the inner PWV area, 1.7 million in greater Durban, and another 1 million split between the Winterveld of Bophuthatswana and KwaNdebele. Such deprivation derives both from the powerlessness of the poor, and the fact that South Africa is increasingly a labour-surplus economy (Tomlinson, 1990, p. 13), with between 25 per cent and 40 per cent of the economically active urban African population formally unemployed. Some two-thirds of all Africans could not afford even a serviced site in 1985, and four-fifths were unable to afford the cheapest formal house (de Vos, 1986b).

The government, clearly alarmed by the dimensions of the problem, has followed the recommendations of the Venter Commission (South Africa, 1984b) in seeking to shift the burden to the private sector. In 1986, it created the South African Housing Trust (SAHT) as a joint government and private enterprise venture, with the latter expected to provide two-thirds of the initial capital. The Trust finances serviced sites and builds basic houses, but on a scale which makes little impact on the total problem. In 1988 a reduction in the budget of the National Housing Commission, forcing it to seek funds from the SAHT, further shifted the financial burden to the private sector. In 1990, the level at which African first-time buyers qualify was considerably reduced to a (household) income of R2,000 per month, apparently because subsidy funds were running out. More positively, the 1990 budget set aside R3 billion to reduce the many backlogs created by apartheid, R2 billion of which has funded a new trust chaired by Jan Steyn, the Urban Foundation chairman, who intends to involve mass-supported organizations in managing the fund.

The SAHT and the Urban Foundation have been trying to create a financial climate which would entice financial institutions into low-income housing areas. The downgrading of banks' and building societies' capital ratios (the percentage owners' capital the institution must maintain against assets), already low by international standards, is seen as a strategy to make low-cost housing finance more profitable, but the risks are considerable (Bond, 1990). The Usury Act has been changed to allow higher administrative costs for small loans to be passed on to borrowers. An insurance scheme to reduce lenders' collateral risk has been introduced. The Foundation is also looking to pension funds and life insurance offices for innovative housing finance. It has itself initiated a pilot group credit scheme in the Western Cape providing unsecured home improvement loans of R1,000–R5,000 to groups whose members stand collateral for each other. A change from monthly subsidy of mortgage costs to a one-off lump sum grant is another possibility: this would have the advantage of removing the present unfairness where more expensive houses are subsidized more, and would enable a squatter to pay a deposit on a serviced site and gradually improve the structure on it.

The above are but a fraction of the many initiatives currently under way in the housing market. The mass housing problem is the subject of vigorous debate and much private sector funded research. The private sector clearly does have an important role to play in the housing market, but there are dangers in taking this role too far down the income scale, which may encourage over-commitment on the part of unsophisticated buyers who may later be unable to cope with changes in both interest rates and their personal circumstances. There are also dangers of unscrupulous building contractors and credit brokers taking advantage of those desperate for low-cost housing (Bond, 1990). With regard to land, Tomlinson (1990, p. 93) is surely right in saying that 'To hold that the private sector can profit from providing the urban poor with serviced land *and* that the land will be affordable is make-believe'.

Tomlinson also identifies the proven inadequacies of housing policies elsewhere in the Third World. Public housing is too expensive to meet the demand, often peripherally located, and inflexible in relation to the investment priorities of the poor. Self-help housing projects can be similarly criticized, and in practice they have often failed to reach the poor: the cost may still be too high, or 'downward raiding' may occur, whereby the poor trade their sites to fulfill even more urgent needs. The *concept* of self-help (or 'sweat equity') must be appropriate in Third World cities, but 'one should strive to open up the housing process so that self-help can be taken more literally' (Tomlinson, 1990, p. 90).

Such opening up requires one thing above all: a supply of well located, affordable land. Where the poor have already found the land for themselves, and the emphasis is on shack upgrading, legitimization is important to create security, together with the incorporation of existing social structures into the planning process (Boaden, 1990). Overall land needs are daunting: allowing for space consumed by roads, schools, parks, commercial and church sites, Wessels (1989) estimated additional land needs in the metropolitan areas at 1,857 square kilometres, or somewhat more than the current area of Greater London. It must be located so that the poor do not live distantly from actual or potential employment. Money currently spent on transport subsidies for Africans whose place in the apartheid city is distant from their jobs − currently more than R1 billion annually − could gradually be redeployed in the provision of better located land. Where in the apartheid city can such land be found?

Post-apartheid planners will inherit a situation in which some land has already been allocated for housing development, and city councils have already bought varying amounts of land contiguous to existing group areas and townships. In the short term at least, the temptation in respect of both housing and land provision may be simply to extend the sectoral patterns of the apartheid city model, albeit without the ethnic zoning which applied before. But as Pirie argues in Chapter 8, there are strong arguments for seeking to 'crack this racial mould' as soon as possible, although there may be costs in doing so. The very association with the past, and all it symbolizes for blacks, is reason enough. The location of such sites may unnecessarily perpetuate or worsen commuting problems for the poor: why, for intance, press ahead with a 'second Soweto' at Rietfontein, 50 kilometres south of Johannesburg, when Johannesburg has vast areas of centrally situated derelict land awaiting development? (Chapter 9).

Overall population densities in South African cities are low, and the post-apartheid city will need to be 'reconstructed' on a more compact basis (Urban Foundation, 1990a). This is likely to involve using accessible areas of waste land, even small ones, for low-income housing and serviced land; buffer strips between former group areas are obvious candidates in many places, although opposition is likely from householders in adjacent suburbs, anxious to preserve their environment and property values. Smaller, scattered townships and in-formal settlements would be more human in scale than the monotonous sprawl of Soweto or Mdantsane, the Winterveld or Botshabelo. They would help, if not to integrate city populations, at least to bring the separate worlds of the apartheid city closer to one another's consciousness.

Space precludes detailed consideration of policies to make such land available for housing the poor, but it would in any case be difficult to improve upon the suggestions made recently by Tomlinson (1990), in a valuable exploration of alternative strategies in the light of relevant international experience. They include two forms of tax: a progressive site value tax such as that used in Taiwan, which increases with property values, would promote smaller land-holdings and increased densities, as well as increasing the share contributed by the wealthy towards the cost of providing and extending municipal services; alongside this, a tax on vacant land has obvious application to post-apartheid cities, as it would help to bring land on to the market earlier, and at lower cost, than would otherwise be the case. Tomlinson also proposes a Land Corporation which could intervene directly in the market and so improve delivery to the poor. Recognizing that the constitution of such a body would be critical if it is to avoid the pitfalls into which autonomous, self-financing and centralized land corporations have fallen in India and elsewhere, Tomlinson proposes a body which depends on borrowing capital, and therefore needs to turn over land as quickly as possible; which would constitute a development arm of local govern-ment, responsible to democratically elected councils; and whose main source of capital would be a central development bank whose conditions for assistance would include the participation of the poor in the design and implementation of projects (Tomlinson, 1990, pp. 111–26, 199–200).

Such an approach to the supply of serviced land has clear implications for the spatial structure of local government itself. Fox, Nel and Reintges have stressed in Chapter 4 the dysfunctional nature of current apartheid structures, which are fragmented not only by the insistence on ethnically defined primary local authorities, but also by the authority exercised by bantustan authorities on the fringes of East London, Durban and Pretoria in particular. Regional Services Councils as presently constituted are not urban planning bodies, and their failure to include bantustan territory means that the limited benefits they bring in terms of upgrading of infrastructure and services cannot reach those whose need is greatest. Single authorities charged with environmental planning, tran-sport and the provision of housing, land and essential services in functional urban regions are urgently needed. Such authorities would receive the bulk of locally generated revenue, from taxes on residents, industry and business, as well as whatever assistance the State was able and willing to provide for local authorities. Allocation of revenue would be according to need rather than derivation.

In the metropolitan regions, and perhaps more widely, second-tier authorities would also be desirable, with more much limited, community derived resources (the equivalent of a parish rate in Britain) and functions which are more appropriately provided at a local or community level, including those serving the cultural needs of locally dominant interest groups. Such second-tier authorities might be permitted to supplement services provided by the city-wide authority, which could enable the affluent to maintain the high quality facilities and amenities to which they have been accustomed, but only if they were prepared to pay extra to do so. Such a system could at least reduce the problem, which Dewar notes in Harare (Chapter 13), of a black middle-class, newly established in former white suburbs, forming a new pressure group to maintain privileged treatment of their suburbs at the expense of those with more pressing needs.

The wider context: rural, regional and economic policies

Our concern in this volume has been with the way in which colonial, segregation and especially apartheid practices have moulded distinctive city forms which are currently experiencing great pressures, and which may soon be open to the imprint of quite different urban management practices. But cities do not exist in a vacuum, and their future is ultimately bound up with wider policies and processes at regional, national, and even international level. Substantive discussion of these issues would require another book, but a brief delineation of their potential significance for post-apartheid cities is relevant.

If allowed to do so, people live where they can work, thus rural development policies affect the nature and directions of urbanization. Any attempt to create economic agricultural units in the bantustans must displace many people from their current limited dependence on inadequate holdings, and it seems unlikely that the economic base of bantustan towns will expand to absorb them. Indeed the reincorporation of the bantustans into South Africa would remove the major functions of their capitals, of which all except Umtata (Transkei) were creations of apartheid. As Drummond and Parnell suggest in Chapter 11, severe hardship could result in these towns unless they are given alternative functions, perhaps as regional capitals. The future of industrial growth points within the bantustans, where even existing firms may well leave when concessions cease to operate, is bleak. Although migration to bantustan towns in unlikely to cease altogether, it is likely to be much exceeded by migration from both urban and rural areas in the bantustans to larger cities.

Another, and even less predictable, element in the demographic equation is the size of the future African population living on the white farms which currently occupy most of South Africa. However, even if some degree of subdivision and transfer to African occupation eventually occurs, it seems unlikely that these areas will absorb significantly more Africans than the number

who currently live and work on white farms. Future levels of African urbanization are thus unlikely to be much affected by what happens to these farms.

It is possible that a post-apartheid government might, following the example
of its predecessor but for different reasons, seek to limit the growth of metropolitan areas by means of a growth centre strategy and the decentralization of
industry. The well-known and costly failures of apartheid decentralization
programmes should ensure that any such policy avoids the pitfalls of too many
growth points in unattractive locations which succeed only in attracting heavily
subsidized, labour intensive industries paying poverty wages, as in Botshabelo.
If there is any scope for constructive regional policy, it rests in the stimulation
of secondary core areas with strong potential for growth (Lemon, 1990a). At
best, however, this is unlikely to do more than close the predicted gap between
the growth rates of such secondary cities and those of the metropolitan areas.

Even this degree of state intervention to modify the urban system may be
undesirable. In much of the Third World, efforts to plan the spatial configuration
of the urban system as an instrument of regional economic policy or to limit the
growth of primate cities have been unsuccessful (Rogerson, 1989b, 1989c;
Tomlinson, 1990). The Kenyan experience, where spatial policies have given
way to an emphasis on the optimal functioning of the urban system to support
accelerated economic development and employment creation, has lessons for
South Africa (Hart and Rogerson, 1989). Commenting on a series of comparative
studies, Rogerson (1989a) notes a close correlation between the most promis-
ing examples of urban management (which included Bogota and Seoul) and
national economic growth rates. Given the essentially Third World character of
urban problems in a post-apartheid South Africa, this suggests that national
urban policy should be geared to the wider goals of national economic planning.

Conclusion

Earlier in this book Gillian Cook commented that the major resource of
greater Cape Town is its people. Ultimately this is true of all South African
cities, and the chief reason for optimism about South Africa's urban future
rests in the hope that decades of minority imposed, ideologically determined
planning may before long be succeeded by an approach centred on all the
people who live, and want to live, in the cities. Post-apartheid planners and
politicians will need to work closely with the communities they serve, trying to
meet their priorities in the face of limited resources and huge needs, and
making the amenities and services of the cities accessible to all. The majority
who are poor have amply and enduringly demonstrated their resilience and
resourcefulness in the face of oppressive urban policies. These qualities must
be harnessed not to mere survival, as so often hitherto, but to the steady
improvement of urban environment and opportunity.

REFERENCES

Abrahams, P. (1963) Return to Goli. *Faber and Faber*, London.

Abu-Lughod, J. (1980) *Rabat: Urban Apartheid in Morocco*, Princeton University Press.

Adam, H. (1971) *Modernizing Racial Domination*, University of California Press, Berkeley.

Arrighi, G. (1970) Labour supplies in historical perspective: A study of the proletarianization of the African peasantry in Rhodesia, *Journal of Development Studies*, Vol. 6, no. 3, pp. 197–234.

Atkinson, A. and Heymans, C. (1988) The future: what do the practitioners think?, in C. Heymans and G. Totemeyer (eds.) op. cit.

Austin, R. (1975) *Racism and Apartheid in Southern Africa: Rhodesia*, UNESCO, Paris.

Batson, E. (1947) Notes on the distribution and density of population in Cape town, *Transactions of the Royal Society of South Africa*, Vol. 31, no. 4, pp. 389–420.

Beavon, K. S. O. (1982) Black townships in South Africa: terra incognita for urban geographers, *South African Geographical Journal*, Vol. 64, no. 1, pp. 1–20.

Bekker, S. B. and Humphries, R. (1985), *From Control to Confusion: The Changing Role of Administration Boards in South Africa, 1971–1983*, Shuter and Shooter, Pietermaritzburg.

Bell, T. (1973) *Industrial Decentralisation in South Africa*, Oxford University Press, Cape Town.

Bettison, D. G. (1950) *A Socio-economic Study of East London, Cape Province with Special Reference to the Non-European peoples*. Unpublished M.A. thesis, University of South Africa, Pretoria.

Bickford-Smith, V. (1984) Keeping your own council, *Studies in the History of Cape Town*, Vol. 5, pp. 188–207.

Black, P. A., Davies, W. J., Wallis, J. L. and McCartan, P. J. (1986) *Industrial Development Strategy for Region D. Final Report, Appendix IV*, Institute of Social and Economic Research, Rhodes University, Grahamstown.

Bley, H. (1971) *South West Africa under German Rule*, Heinemann, London.

Boaden, B. (1990) The myths and the realities of shack upgrading, *Urban Forum*, Vol. 1, no. 1, pp. 75–84.

Bond, P. (1990) Township housing and South Africa's 'financial explosion': the theory and practice of financial capital in Alexandra, *Urban Forum*, Vol. 1, no. 2, pp. 39–67.
Bonner, P. L. (1989) The politics of black squatter movements on the Rand, 1944–1952, *Review of Radical History*, Vol. 46/7, no. 3, pp. 89–116.
Boraine, A. (1990) Managing the urban crisis, 1986–1989: the role of the National Security Management System, in G. Moss and I. Obery (eds.) *South Africa Contemporary Analysis: South African Review Five*, Hans Zell, London.
Brindley, M. (1976) *Western Coloured Township*, Ravan, Johannesburg.
Budlender, D. (1984) *Incorporation and Exclusion: Recent Developments in Labour Law and Influx Control*, South African Institute of Race Relations, Johannesburg.
Burton, I. (1963) A restatement of the dispersed city hypothesis, *Annals of the Association of American Geographers*, Vol. 53, no. 3, pp. 285–89.
Cabinet Committee on Housing (1988) Proposals Towards the Establishment of a Uniform Housing Policy for South West Africa/Namibia: final report submitted to cabinet of the government of national unity by the cabinet committee on housing. Unpublished report, Windhoek, August.
Chanaiwa, D. (1981) *The Occupation of Southern Rhodesia: A Study of Economic Imperialism*, East African Publishing House, Nairobi.
Chapman, R. A. and Cristallides, D. (1983) *Some Theoretical Perceptions, Major Policy Issues and Principles Concerning Urban Transportation Policy in South Africa*. Project Report, Urban Problems Research Unit, University of Cape Town.
Christie, R. (1987) *Group Areas and Property Market Economics*, South African Institute of Race Relations, Johannesburg.
Christopher, A. J. (1970) Salisbury 1900: the study of a pioneer town, *Journal for Geography*, Vol. 3, no. 7, pp. 757–66.
Christopher, A. J. (1983) From Flint to Soweto: reflections on the colonial origins of the apartheid city, *Area*, Vol. 15, no. 2, pp. 145–9.
Christopher, A. J. (1984) *Colonial Africa*, Croom Helm, London.
Christopher, A. J. (1987a) Race and residence in colonial Port Elizabeth, *South African Geographical Journal*, Vol. 69, no. 1, pp. 3–20.
Christopher, A. J. (1987b) Apartheid planning in South Africa: the case of Port Elizabeth, *Geographical Journal*, Vol. 153, no. 2, pp. 195–204.
Christopher, A. J. (1988) Segregation and population distribution in Port Elizabeth, *Contree: Journal for South African Urban and Regional History*, No. 24, pp. 5–12.
Christopher, A. J. (1989a) Spatial variations in the application of residential segregation in South African cities, *Geoforum*, Vol. 20, no. 3, 253–67.
Christopher, A. J. (1989b) Apartheid within apartheid: an assessment of official intrablack segregation on the Witwatersrand, South Africa, *The Professional Geographer*, Vol. 41, no. 3, pp. 328–35.
Christopher, A. J. (1990) Apartheid and urban segregation levels in South Africa, *Urban Studies*, Vol. 27, no. 3, pp. 417–36.
Christopher, A. J. (1991) Changing patterns of group area proclamation in South Africa, 1950–1989, *Political Geography Quarterly*, Vol. 10 (forthcoming).
City Engineer's Department (1983) People, employment and land in the eighties, Metropolitan Transport Planning Branch, Cape Town.
City of Cape Town (1986) Demarcation of a regional services council for Greater Cape Town, City Council memorandum.
City of Cape Town (1987) White Paper on Urbanisation: comments from the city of Cape Town, City Planner's Department.
City of Cape Town (1988) Atlantis: its role in the economic development of metropolitan Cape Town, Town Planning Branch.
City of Harare (1987) Minute of His Worship the Mayor, City Council.
City of Pietermaritzburg (1924) *Pietermaritzburg Corporation Yearbook 1923/24*.
City of Pietermaritzburg (1961) *Pietermaritzburg Corporation Yearbook 1960/61*.
Clark, C., Cameron, B., Naude, A., Dehlen, G., and Lipman, V. (1988) *Trends,*

Patterns and Forecasts of Black Commuting in South Africa, Technical Report RT 120, CSIR, Pretoria.

Cloete, J. J. N. (1986) *Towns and Cities: their Government and Administration*, J. L. van Schaik, Pretoria.

Cobbett, W. (1987) Industrial decentralization and exploitation: the case of Botshabelo, *South African Labour Bulletin*, Vol. 12, no. 3, pp. 1–13.

Cobbett, W. and Nakedi, B. (1988) The flight of the Herschelites: ethnic nationalism and land dispossession, in W. Cobbett and R. Cohen (eds.) *Popular Struggles in South Africa*, James Currey, London.

Cokwana, M. M. (1988) A close look at tenure in cities, in C. R. Cross and R. J. Haines (eds.) *Towards Freehold? Options for Land Development in South Africa's Black Rural Areas*, Juta, Cape Town.

Cole, J. (1986) When your life is bitter you do something: women and squatting in the Western Cape, in D. Kaplan (ed.) *South African Research Papers*, Department of Economic History, University of Cape Town.

Cole, J. (1987) *Crossroads: The Politics of Reform and Repression*, Ravan Press, Johannesburg.

Collinge, J. (1990) Sharing the land, *Work in Progress*, No. 66, pp. 15–19.

Cook, G. P. (1986) Khayelitsha – policy change or crisis response?, *Transactions of the Institute of British Geographers*, Vol. 11, no. 1, pp. 57–66.

Corbett, P. J. (1980) *Housing Conditions in Chatsworth, Durban*, Institute of Social and Economic Research, University of Durban-Westville.

Cowley, J. W. (1985a) Mafikeng to Mmabatho: village to capital city. Paper given at the Society for Geography Conference, Stellenbosch.

Cowley, J. W. (1985b) *Bophuthatswana: The Land and the People*, University Bookshop Publishers, Mafikeng.

Cowley, J. W. (1990) Personal communication, Department of Geography, University of Bophuthatswana, and adviser for the Bophuthatswana Population Census.

Crankshaw, O. (1990) Apartheid and economic growth: craft unions, capital and the state in the South African building industry, 1945–1975, *Journal of Southern African Studies*, Vol. 16, no. 3, pp. 503–26.

Crankshaw, O. and Hart, T. (1990) The roots of homelessness: causes of squatting in the Vlakfontein settlement south of Johannesburg, *South African Geographical Journal*, Vol. 72, no. 2, pp. 65–70.

Cross, C. R., Mtimkulu, P., Napier, C. and van der Merwe, L. (1988) Conflict and Violence in Pietermaritzburg: the development factor, *Working Documents in African Politics*, no. 2/1988, University of South Africa, Pretoria.

Cummings, S. D. (1991) Post-colonial urban residential change in Zimbabwe: a case study, in R. B. Potter and A. T. Salan (eds.) *Cities and Development in the Third World*, Mansell, London.

Cuthbertson, G. C. (1979) A new town in Uitflugt, *Studies in the History of Cape Town*, Vol. 1, pp. 93–106.

Dale, P. (1969) The tale of two towns (Mafeking and Gaberone) and the political modernisation of Botswana, *South African Institute of Public Administration*, Vol. 4, no. 1, pp. 130–44.

Darden, J. T. (1989) Blacks and other racial minorities: the significance of color in inequality, *Urban Geography*, Vol. 10, no. 6, pp. 562–77.

Datakonsult (1989) *Population 2000*, Pretoria.

Davenport, R. (1971) *The Beginnings of Urban Segregation in South Africa: the Natives (Urban Areas) Act of 1923 and its Background*, Institute of Social and Economic Research. Occasional Paper no. 15, Rhodes University, Grahamstown.

Davies, R. J. (1981) The spatial formation of the South African city, *GeoJournal*, Supplementary Issue 2, pp. 59–72.

Davies, R. J. (1986) Government policy and the urban settlement system of South Africa, in L. S. Bourne, B. Cori and K. Dziewonski (eds.) *Progress in Settlement*

Systems Geography, Franco Angeli, Milan.

Davies, R. J. (1988) Adaptive or Structural Transformation: the Case of Harare, Zimbabwe. Unpublished seminar paper, Department of Environmental and Geographical Science, University of Cape Town.

Davies, R. J. and Dewar, N. (1989) Adaptive or structural transformation? The case of Harare, Zimbabwe, housing system, *Social Dynamics*, Vol. 15, no. 1, pp. 46−60.

Davies, W. J. (1971) *Patterns of Non-White Population Distribution in Port Elizabeth with Special Reference to the Application of the Group Areas Act*, Series B, Special Publication 1, Institute of Planning Research, University of Port Elizabeth.

De Coning, C., Fick, J. and Olivier, N. (1987) Residential settlement patterns: a pilot study of socio-political perceptions in grey areas of Johannesburg, *South Africa International*, Vol. 18, no. 2, pp. 121−37.

De Klerk, W. A. (1975) *The Puritans in Africa: a story of Afrikanerdom*, Rex Collings, London.

De Vos, T. J. (1986a) Housing shortages and surpluses and their bearing on the Group Areas Act, in M. Rajah (ed.) *The Future of Residential Group Areas*, School of Business Leadership, University of South Africa, Pretoria.

De Vos, T. J. (1986b) Financing Low-cost Housing. Paper presented at a seminar organized by the South African Institute of Building, CSIR, Pretoria, August (cited by Tomlinson [1990], op. cit).

Dewar, D. (1980) Urban development and an open society, in D. Walker (ed.) *Cape Town: An Open City in an Open Society*, University of Cape Town, Cape Town.

Dewar, D. (1986) City planning and urbanisation strategies: meeting the challenge, in C. Heymans and G. Totemeyer (eds.), op. cit.

Dewar, D. and Watson, V. (1990) *An Overview of Development Problems in the Cape Metropolitan Area*, Urban Problems Research Unit, University of Cape Town and the Urban Foundation, Cape Town.

Dewar, N. (1988) *From Salisbury to Harare: The Geography of Public Authority Finance and Practice under Changing Ideological Circumstances*. Unpublished PhD thesis, University of Cape Town.

Drummond, J. H. and Manson, A. H. (1990) The evolution and contemporary significance of the Bophuthatswana (South Africa)-Botswana border, in D. Rumley and J. Minghi (eds.) *The Geography of Border Landscapes*, Routledge, London.

Duncan, O. D. and Davis, B. (1953) *The Chicago Urban Analysis Project*, University of Chicago.

Duncan, O. D. and Duncan, B. (1955) A methodological analysis of segregation indexes, *American Sociological Review*, Vol 20, no. 2, pp. 210−17.

Duncan, S. (1985) Social and practical problems resulting from the law relating to urban Africans, *Acta Juridica 1984*, Juta for the Faculty of Law, University of Cape Town, pp. 247−54.

East London Municipality (1949−50) *Location Enquiry (Welsh Commission)* (Vol. 50/665/17).

Eixab, S. (1986) Housing in Katatura, in C. von Garnier (ed.) op. cit.

Elder, G. (1990), The grey dawn of South African racial residential integration, *GeoJournal*, Vol. 22, no. 3, pp. 261−6.

Elphick, R. H. and Giliomee, H. (1979) *The Shaping of South South African Society: 1652−1820*, Maskew Millar, Cape Town.

Els, W. (1990) Personal communication, 17 May.

Fair, T. J. D. and Davies, R. J. (1976) Constrained urbanization: White South Africa and Black Africa compared, in B. J. L. Berry (ed.) *Urbanization and Counterurbanization*, Sage, Beverly Hills.

Fair, T. J. D. and Muller, J. G. (1981) The Johannesburg metropolitan area, in M. Pacione (ed.) *Urban Problems and Planning in the Developing World*, Croom Helm, London.

Fick, J., de Coning, C. and Olivier, N. (1988), Ethnicity and residential patterning in a divided society: a case study of Mayfair in Johannesburg, *South Africa International*, Vol. 19, no. 1, pp. 1–27.

Fourie, E. M. (1985) The Effect of a Long Distance to Work on the Daily Activities of Black Commuters: A Case Study. Paper given at the Society for Geography Conference, Stellenbosch, 1–6 July.

Fox, R. C. (1990) Ethnic Regions and the Political Process in Africa: A Preliminary Assessment. Unpublished seminar paper, Institute of Social and Economic Research, Rhodes University, 29 May.

Frankel, P. (1979) Municipal transformation in Soweto: race, politics and maladministration in black Johannesburg, *African Studies Review*, Vol. 22, no. 1, pp. 49–63.

French, K. (1982) Squatters in the forties, *Africa Perspective*, Vol. 21, pp. 2–8.

Ginsberg, N. S. (1961) The dispersed metropolis: the case of Okayama, *The Toshi Mundai*, Vol. 52, pp. 631–40.

Gordimer, N. (1962) *A World of Strangers*, Penguin, Harmondsworth (first published by Gollancz in 1958).

Gordon, T. J. (1978) *Mdantsane: City, Satellite or Suburb*. Unpublished M. A. thesis, Rhodes University, Grahamstown.

Grahamstown Rural Committee (1988–90) Newsletters for December 1988, June 1989, March and May 1990.

Greenberg, S. and Giliomee, H. (1985) Managing influx control from the rural end: the black homelands and the underbelly of privilege, in H. Giliomee and L. Schlemmer (eds.) *Up Against the Fences: Passes and Privilege in South Africa*, David Philip, Cape Town.

Grinkler, D. (1986) *Inside Soweto*, Eastern Enterprises, Johannesburg.

Hart, D. M. (1988) District Six – political manipulation of urban space, *Urban Geography*, Vol. 9, no. 6, pp. 603–28.

Hart, D. M. and Pirie, G. H. (1984) The sight and soul of Sophiatown, *Geographical Review*, Vol. 74, no. 1, 38–47.

Hart, D. M. and Rogerson, C. M. (1989) Urban management in Kenya: South African policy issues, *South African Geographical Journal*, Vol. 71, no. 3, pp. 192–200.

Hart, G. (1989) On grey areas, *South African Geographical Journal*, Vol. 71, no. 2, pp. 81–8.

Hart, G. H. T. (1990) Susceptibility of Johannesburg's Housing Market to Political Unrest. Unpublished paper to the International Sociological Congress, Madrid, July.

Harvey, S. D. (1987) Black residential mobility in a post-independence Zimbabwean city, in G. J. Williams and A. P. Wood. (eds.) *Geographical Perspectives on Development in Southern Africa*, papers from the Regional Conference of the Commonwealth Geographical Bureau, Lusaka.

Haswell, R. F. (1979) South African towns on European plans, *Geographical Magazine*, Vol. 51, no. 10, pp. 686–94.

Haswell, R. F. (1985) Indian townscape features in Pietermaritzburg, *Natalia*, Vol. 15, pp. 23–8.

Hattingh, P. S. (1975) Nedersetting en Verstedeliking, in J. H. Moolman and G. M. E. Leistner (eds.) *Bophuthatswana: hulpbronne en ontwikkeling*, Africa Institute, Pretoria.

Hellman, E. (1949) Urban Areas, in E. Hellman (ed.) *Handbook on Race Relations in South Africa*, South African Institute of Race Relations, Johannesburg.

Hendler, P. (1988) *Urban Policy and Housing*, South African Institute of Race Relations, Johannesburg.

Hendler, P. (1989) *Politics on the Home Front*, South African Institute of Race Relations, Johannesburg.

Heymans, C. and Totemeyer, G. (eds.) (1988) *Government by the People?*, Juta, Cape Town.

Hindson, D. (1987) *Pass Controls and the Urban African Proletariat*, Ravan, Johannesburg.

Hindson, D. and Lacey, M. (1983) Influx control and labour allocation: policy and practice since the Riekert Commission, *South African Review One*, Ravan, Johannesburg.

Hooper, C. (1960) *Brief Authority*, Collins, London.

Horrell, M. (1956) *The Group Areas Act: Its Effects on Human Beings*, South African Institute of Race Relations, Johannesburg.

Humphries, R. (1989) A tale of two squirrels: the 1988 local government elections and their implications, in Centre for Policy Studies, *South Africa at the end of the Eighties*, Centre for Policy Studies, University of the Witwatersrand, Johannesburg.

Jochelson, K. (1990) Reform, repression and resistance in South Africa: a case study of Alexandra township, 1979−1989, *Journal of Southern African Studies*, Vol. 16, no. 1, pp. 1−32.

Junod, H. P. (1955) The bantu population of Pretoria, in S. P. Engelbrecht (ed.) *Pretoria 1855−1955*, Pretoria City Council.

Kagan, N. (1978) African settlements in the Johannesburg area, 1903−1923. Unpublished M.A. dissertation, University of the Witwatersrand, Johannesburg.

Kay, G. and Smout, M. A. H. (eds.) (1977) *Salisbury: A Geographical Survey of the Capital of Rhodesia*, Hodder and Stoughton, London.

King, A. D. (1976) *Colonial Urban Development: Culture, Social Power and Environment*, Routledge and Kegan Paul, London.

King, A. D. (1985) Colonial cities: global pivots of change, in G. Telkamp and R. Ross (eds.) *Colonial Cities*, University of Leiden Press.

Koch, E. (1983) Without visible means of subsistence: slumyard culture in Johannesburg 1918−1940, in B. Bozzoli (ed.) *Town and Countryside in the Transvaal*, Ravan, Johannesburg.

Krige, D. S. (1988a) Die Transformasie van die Suid-Afrikaanse Stad, *Town and Regional Planning Research Publication* no. 10, University of the Orange Free State, Bloemfontein.

Krige, D. S. (1988b) *Afsonderlike Ontwikkeling as ruimtelike beplanningstrategie: 'n toepassing op die Bloemfontein-Botshabelo-Thaba Nchu-streek*. Unpublished D Phil thesis, University of the Orange Free State, Bloemfontein.

Krige, D. S. (1990) Apartheidsbeplanning in die Bloemfontein-Botshabelo-Thaba Nchu-Streek, *South African Geographer*, Vol 17, no. 1/2, pp. 14−34.

Kuper, L., Watts, H. and Davies, R. J. (1958) *Durban: A Study in Racial Ecology*, Jonathan Cape, London.

Laband, J. and Haswell, R. H. (eds.) (1988) *Pietermaritzburg 1838−1988: A New Portrait of an African City*, University of Natal Press/Shooter & Shuter, Pietermaritzburg.

Legge, P. (1984) Personal communication (primary school headmistress in the Stadt).

Lelyveld, J. (1986) *Move Your Shadow: South Africa, black and white*, Michael Joseph, London.

Lemon, A. (1982) Migrant labour and frontier commuters: reorganizing South Africa's black labour supply, in D. M. Smith (ed.) *Living under Apartheid: aspects of urbanization and social change in South Africa*, Allen and Unwin, London.

Lemon, A. (1987) *Apartheid in Transition*, Gower, Aldershot.

Lemon, A. (1990a) Urban policy options to prevent rural and urban overcrowding: the size and location of urban settlement. Confidential paper given at the Newick Park Conference, Sussex, 26−29 July 1989, substantially reprinted in *Towards an Urbanization and Housing Policy for Post-Apartheid South Africa*, Christian Research and Education Information for Democracy (CREID), Johannesburg.

Lemon, A. (1990b) Geographical issues and outcomes in the South African 'general' election, *South African Geographical Journal*. Vol. 72, no. 2, pp. 54−64.

Lemon, A. (1991) Restructuring the local state in South Africa: Regional Services Councils, redistribution and legitimacy, in D. Drakakis-Smith (ed.) *Urban and Regional Change in South Africa*, Routledge, London.

Lewis, P. R. B. (1966) A 'city' within a city − the creation of Soweto', *South African Geographical Journal*, Vol. 48, no. 1, pp. 45−85.

LHA (1990) *The Black Demographic Handbook for the 1990s*, LHA Management Consultants, Pretoria.

Lodge, T. (1981) The destruction of Sophiatown. *Journal of Modern African Studies*, Vol. 19, no. 2, pp. 107–32.

Louw, A. D. (1986) The attitudes of white owners of fixed property in Mafikeng towards local intergroup relations four years after incorporation into Bophuthatswana, *Plural Societies*. Vol. 16, no. 2, pp. 189–200.

Lowder, S. (1986) *Inside Third World Cities*, Croom Helm, London.

Maasdorp, G. and Humphreys, A. S. B. (1975) *From Shantytown to Township*, Juta, Cape Town.

Maasdorp, G. and Pillay, N. (1975) Memoranda on behalf of the Indian traders of Bethal (regarding their proposed removal to a shopping centre in Milan Park) and Piet Retief (regarding their proposed removal to a shopping centre at Kempville township), Department of Economics, University of Natal, Durban.

Maasdorp, G. and Pillay, N. (1977) *Urban Relocation and Racial Segregation: the case of Indian South Africans*. Research Monograph, Department of Economics, University of Natal, Durban.

Mabin, A. (1986) Labour, capital, class struggle and the origins of residential segregation in Kimberley, 1880–1920, *Journal of Historical Geography*, Vol. 12, no. 1, pp. 4–26.

Mabin, A. (1989) Struggle for the city: urbanisation and political strategies of the South African State, *Social Dynamics*, Vol. 15, no. 1, pp. 1–28.

Mandy, N. (1984) *A City Divided*, Macmillan, Johannesburg.

Mashabela, H. (1988) *Townships of the PWV*, South African Institute of Race Relations, Johannesburg.

Mashabela, H. (1990) *Mekhuhu: Urban African Cities of the Future*, South African Institute of Race Relations, Johannesburg.

Mashile, G. G. and Pirie, G. H. (1977) Aspects of housing allocation in Soweto, *South African Geographical Journal*, Vol 59, no. 2, pp. 139–49.

Massey, D. S. and Denton, N. A. (1987) Trends in the residential structure of Blacks, Hispanics and Asians: 1970–1980, *American Sociological Review*, Vol. 52, no. 6, pp. 802–25.

Mather, C. (1985) Racial Zoning in Kimberley, 1951–1959. Unpublished seminar paper, Department of Geography, University of the Witwatersrand, Johannesburg.

Mather, C. (1987) Residential segregation and Johannesburg's 'locations in the sky', *South African Geographical Journal*, Vol. 69, no. 2, pp. 119–28.

Mather, C. and Parnell, S. M. (1990) Upgrading the matchboxes: urban renewal in Soweto, 1976–1986, in D. Drakakis-Smith (ed.) *Economic Growth and Urbanization in Developing Areas*, Routledge, London.

Maylam, P. (1990) The rise and decline of urban apartheid in South Africa, *African Affairs*, Vol. 89, no. 354, pp. 57–84.

McGee, T. G. (1971) *The Urbanization Process in the Third World: explorations in search of a theory*, Bell and Hyman, London.

McGrath, M. D. (1989) Income distribution, expenditure patterns and poverty. Paper prepared for *The Durban Functional Region: planning for the 21st century*, Report no. 1, Tongaat-Hulett Properties Ltd.

Meer, F. (1971) Indian people: current trends and policies, in *South Africa's Minorities*, SPRO-CAS Publication no. 2, pp. 13–32.

Melber, H. (1979) *Schule und Kolonialismus: Das Formale Erziehungswesen Namibias*, Institut fur Afrika-Kunde, Hamburg.

Minkley, G. (1985) *'To Keep in Your Hearts': The I. I. C. U. Class Formation and Popular Struggle*. Unpublished BA Honours dissertation, University of Cape Town.

Moadira, M. I. (1984) Urban Renewal in Montshiwa Township. Unpublished BA Ed. project, Department of Geography, University of Bophuthatswana, Mmabatho.

Morrill, R. (1965) The negro ghetto: problems and alternatives, *Geographical Review*, Vol. 55, no. 3, pp. 339–61.

Morris, M. (1977) *Apartheid, Agriculture and the State: the Farm Labour Question*,

Working Paper no. 8, South African Labour and Development Research Unit, University of Cape Town.

Morris, N. (1983) *Black Commuting in Pretoria: Population and Employment Figures*. Special Report BCP 1, National Institute for Transport and Road Research, CSIR, Pretoria.

Morris, P. (1981a) *Soweto*, Urban Foundation, Johannesburg.

Morris, P. (1981b) *A History of Black Housing in South Africa*, South Africa Foundation, Johannesburg.

Motala, M. M. (1961) The Group Areas Act and the Effects of the Proclamations made Thereunder in Pietermaritzburg and Environs. Paper given at the Group Areas Conference, Pietermaritzburg, January.

Muller, A. M. (1988) *Housing as part of the Process of Change in Namibia*. Unpublished M. Phil. thesis, University of Newcastle upon Tyne.

Municipality of Windhoek (1987) *Resultate van Bevolkingsopname: Mei 1985*, Town Planning Section, City Engineer's Department.

Natal Tower and Regional Planning Commission (1973) Pietermaritzburg Durban Region: Regional Guide Plan, NTRT Reports, vol. 24.

National Building and Investment Corporation of SWA/Namibia (1986) *Annual Review 1985*, Windhoek.

National Building and Investment Corporation of SWA/Namibia (1989) *Proposed Five Year Construction Targets for NBIC 1990−1994*, Windhoek.

Nel, E. L. (1989) The Evolution of Racial Residential Segregation in East London: 1849−1949. Paper given at the Society for Geography Conference, University of Pretoria, 2−6 July.

Nel, E. L. (1990) *The Spatial Planning of Racial Residential Segregation in East London, 1948−1973*. Unpublished M. A. thesis, University of the Witwatersrand, Johannesburg.

Nel, J. G. (1988) *Die geografiese impak van die Wet op Groepsgebiede en verwante wetgewing op Port Elizabeth*, Institute of Planning Research, University of Port Elizabeth.

Niddrie, D. (1988) Into the valley of death, *Work in Progress*, Vol. 52, pp. 6−15.

O'Callaghan, M. (1977a) *Namibia: The Effects of Apartheid on Culture and Education*, UNESCO, Paris.

O'Callaghan, M. (1977b) *Southern Rhodesia: the effects of a conquest society on education, culture and information*, UNESCO, Paris.

Olivier, J. J. and Booysen, J. J. (1983) Some impacts of black commuting in Pretoria, *South African Geographical Journal*, Vol. 65, no. 2, pp. 123−34.

Olivier, J. J. and Hattingh, P. S. (1985) Die Suid-Afrikaanse stad as funksioneel-ruimtelike sisteem met besondere verwysing na Pretoria, *RSA 2000*, Vol. 7, no. 1, pp. 7−18.

Olivier, M. J. (1953) Die naturel in Wes-Kaapland, *Journal of Racial Affairs*, Vol. 4, no. 2, pp. 1−12.

Palley, C. (1966) *The Constitutional History and Law of Southern Rhodesia, 1888−1965*, with Reference to Imperial Control, Clarendon Press, Oxford.

Parnell, S. (1986) From Mafeking to Mafikeng: the transformation of a South African town, *GeoJournal*, Vol. 12, no. 2, pp. 203−10

Parnell, S. M. (1988a) Public housing for whites as a device of residential segregation, 1934−1939, *Urban Geography*, Vol. 9, no. 6, pp. 584−602.

Parnell, S. M. (1988b) Racial segregation in Johannesburg, the Slums Act, 1934−1939, *South African Geographical Journal*, Vol. 70, no. 2, pp. 112−126.

Parnell, S. M. (1989a) Creating a racially divided society: state housing policy, 1920−1955, *Government and Policy: Environment and Planning* C, Vol. 7, no. 3, pp. 261−72.

Parnell, S. M. (1989b) The Ideology of African Home Ownership: the case of Dube, Soweto. Paper given at the Historical Geography Conference, Jerusalem, August.

Parnell, S. M. and Webber, H. (1990) The privatization of black secondary schooling and the desegregation of Johannesburg, *GeoJournal*, Vol. 22, no. 3, pp. 267–73.

Parsons, N. (1982) *A New History of Southern Africa*, Macmillan, London.

Peel, H. (1988) Sobantu, in J. Laband and R. H. Haswell (eds.) op. cit.

Pendleton, W. C. (1974) *Katutura: a Place Where we do Not Stay*, San Diego University Press.

Phimister, I. (1975) Peasant production and underdevelopment in Southern Rhodesia, *African Affairs*, Vol. 13, no. 291, pp. 217–28.

Pickard-Cambridge, C. (1988a) *The Greying of Johannesburg*, South African Institute of Race Relations, Johannesburg.

Pickard-Cambridge, C. (1988b) *Sharing the Cities: Residential Desegregation in Harare, Windhoek and Mafikeng*, South African Institute of Race Relations, Johannesburg.

Pirie, G. H. (1982) Mostly 'Jubek': urbanism in South African English literature?, *South African Geographical Journal*, Vol. 64, no. 1, pp. 63–7.

Pirie, G. H. (1983) Urban bus boycott in Alexandra township, 1957, *African Studies*, Vol. 42, no. 2, pp. 67–77.

Pirie, G. H. (1984a) Letters, words, worlds: the naming of Soweto, *African Studies*, Vol. 43, no. 1, pp. 43–51.

Pirie, G. H. (1984b) Ethno-linguistic zoning in South African black townships, *Area*, Vol. 16, no. 4, pp. 291–8.

Pirie, G. H. (1984c) Race zoning in South Africa: board, court, parliament, public, *Political Geography Quarterly*, Vol. 3, no. 3, pp. 207–21.

Pirie, G. H. (1986) Johannesburg transport, 1905–1945: African capitulation and resistance, *Journal of Historical Geography*, Vol. 12, no. 1, pp. 41–55.

Pirie, G. H. (1987) Deconsecrating a 'Holy Cow': reforming the Group Areas Act, *South African Review Four*, Ravan, Johannesburg.

Pirie, G. H. (1988) Housing essential service workers in Johannesburg: locational constraint and conflict, *Urban Geography*, Vol. 9, no. 6, pp. 568–83.

Pirie, G. H. and da Silva, M. (1986) Hostels for African migrants in greater Johannesburg, *GeoJournal*, Vol. 12, no. 2, pp. 173–80.

Pirie, G. H. and Hart, D. M. (1985) The transformation of Johannesburg's black western areas, *Journal of Urban History*, Vol. 11, pp. 387–410.

Platzky, L. and Walker, C. (1985) *The Surplus People: Forced Removals in South Africa*, Raven, Johannesburg.

Plewman, R. P. (Chairman) (1958) *Report of the Urban Affairs Commission*, Government Printer, Salisbury.

Quenet, V. (Chairman) (1967) *Report of the Commission of Inquiry into Racial Discrimination*, Government Printer, Salisbury.

Randall, P. (1973) *From Coolie Location to Group Area*, South African Institute of Race Relations, Johannesburg.

Reintges, C. (1989) Orderly Urbanization: the Case of Duncan Village. Paper given at the Society for Geography conference, University of Pretoria, 2–6 July.

Ribnick, A. (1987) *Black Entry into the Former White Housing Market in Harare*. Unpublished BA Honours dissertation, Department of Environmental and Geographical Science, University of Cape Town.

Rich, P. B. (1978) Ministering to the White man's needs: the development of urban segregation in South Africa 1913–23, *African Studies*, Vol. 37, no. 2, pp. 177–91.

Riordan, R. (1988a) Life on the low road, *Optima*, Vol. 36, no. 2, pp. 58–67.

Riordan, R. (1988b) Ukuhleleleke of Port Elizabeth, *Monitor*, No. 1, pp. 2–16.

Rip, M. R. and Hunter, J. M. (1990) The community peri-natal health care system of urban Cape Town – (2) Geographical patterns, *Social Science Medicine*, Vol. 30, no. 1, pp. 119–30.

Roberts, B. (1984) *Kimberley: Turbulent City*, Ravan, Johannesburg.

Rogerson, C. M. (1989a) Managing urban growth in South Africa: learning from the international experience, *South African Geographical Journal*, Vol. 71, no. 3, pp. 129–33.

Rogerson, C. M. (1989b) Rethinking national urban policies: lessons for South Africa, *South African Geographical Journal*, Vol. 71, no. 3, pp. 134—41.

Rogerson, C. M. (1989c) Managing the decolonizing city in Southern Africa, *South African Geographical Journal*, Vol. 71, no. 3, pp. 201—208.

Rogerson, C. M. (1990) Drinking apartheid and the removal of beerhalls in Johannesburg, 1938—1962, in C. Ambler and J. Crush (eds.), *Drinking in Africa: Alcohol, Social Control and the State in Eastern and Southern Africa*, James Currey, London.

Rule, S. P. (1988) Racial residential integration in Bertrams, Johannesburg, *South African Geographical Journal*, Vol. 70, no. 1, pp. 69—72.

Rule, S. P. (1989) The emergence of racially mixed residential areas in Johannesburg: demise of the apartheid city?, *Geographical Journal*, Vol. 155, no. 2, pp. 196—203.

Saff, G. (1990) The probable effects of the introduction of Free Settlement Areas in Johannesburg, *Urban Forum*, Vol. 1, no. 1, pp. 5—27.

Sampson, A. (1987) *Tycoons, Revolutionaries and Apartheid*, Hodder and Stoughton, London.

Savage, M. (1986) The imposition of pass laws on the African population in South Africa 1916—1984, *African Affairs*, Vol. 85, no. 339, pp. 181—205.

Schoeman, K. (1980) *Bloemfontein: die ontstaan van 'n stad, 1846—1946*, Human and Rousseau, Cape Town.

Schoombee, J. J. (1985) Group Areas legislation in the political control of ownership and occupation of land, *Acta Juridica 1984*, Juta for the Faculty of Law, University of Cape Town, pp. 77—118.

Schlemmer, L. (1986) Apartheid in transition: the collapse of racial zoning, *Indicator SA*, Vol. 4, no. 2, pp. 32—6.

Schlemmer, L. S. and Stack, S. L. (1989) *Black, White and Shades of Grey: A Study of Responses to Residential Segregation in the Pretoria-Witwatersrand Region*, Centre for Policy Studies, University of the Witwatersrand, Johannesburg.

Scott, P. (1955) Cape Town — a multiracial city, *Geographical Journal*, Vol. 121, no. 2, pp. 149—57.

SECOSAF (1985) *Manual on the implementation of the Regional Development Incentives introduced in 1 April 1982*, Pretoria.

Shannon, H. A. (1937) Urbanization, 1904—1936, *South African Journal of Economics*, Vol. 5, no. 2, pp. 164—90.

Shorten, J. R. (1963) *The Johannesburg Saga*, J. Shorten (Pty) Ltd., Johannesburg.

Simkins, C. (1990) People power, *Leadership*, Vol. 9, no. 5, pp. 59—60.

Simon, D. (1983) The evolution of Windhoek, in C. Saunders (ed.) *Perspectives on Namibia: Past and Present*, Paper 4, Centre for African Studies, University of Cape Town.

Simon, D. (1984a) Third World colonial cities in context: conceptual and theoretical issues with particular respect to Africa, *Progress in Human Geography*, Vol. 8, no. 4, pp. 493—524.

Simon, D. (1984b) Urban poverty, informal sector activity and intersectoral linkages: evidence from Windhoek, Namibia, *Development and Change*, Vol. 15, no. 4, pp. 551—576.

Simon, D. (1985) Decolonisation and local government in Namibia: the neo-apartheid experiment, 1977—83, *Journal of Modern African Studies*, Vol. 23, no. 3, pp. 507—26.

Simon, D. (1986) Desegregation in Namibia: the demise of urban apartheid?, *Geoforum*, Vol. 17, no. 2, pp. 289—307.

Simon, D. (1988a) Urban squatting, low income housing and politics in Namibia on the eve of independence, in R. A. Obudho and C. C. Mhlanga (eds.) *Slum and Squatter Settlements in Sub-Saharan Africa: towards a planning strategy*, Praeger, New York.

Simon, D. (1988b) Katutura: symbol stadtischer apartheid, in H. Melber (ed.) *Katutura: Alltag in Ghetto*, Informationsstelle Sudliches Afrika, Bonn.

Simon, D. (1989a) Crisis and change in South Africa: implications for the apartheid city, *Transactions of the Institute of British Geographers*, Vol. 14, no. 2, pp. 189—206.

Simon, D. (1989b) Colonial cities, postcolonial Africa and the world economy: a reinterpretation, *International Journal of Urban and Regional Research*, Vol. 13, no. 1, pp. 68—91.

Smit, P. and Booysen, J. J. (1977) *Urbanization in the Homelands — a new dimension in the urbanization process of the black population of South Africa?*, Institute for Plural Societies, University of Pretoria, Monograph Series on Inter-group Relations, no. 3.

Smith, D. M. (1991) *Urbanization in Contemporary South Africa: the Apartheid City and Beyond*, Unwin Hyman, London.

Smuts, J. C. (1942) *The Basis of Trusteeship*, New Africa Pamphlet 2, South African Institute of Race Relations, Johannesburg.

South Africa (1921) *Report of the Asiatic Inquiry Commission*, UG 4, Government Printer, Pretoria.

South Africa (1941) *Report of the First Indian Penetration Commission*, UG 39, Government Printer, Pretoria.

South Africa (1943) *Report of the Second Indian Penetration Commission*, UG 21, Government Printer, Pretoria.

South Africa (1948a) *Report of the Native Laws Commission, 1946—1948*, UG 28/1948, Pretoria.

South Africa (1948b) *Report of the Judicial Commission on Native Affairs in Durban*, Government Printer, Pretoria.

South Africa (1979) *Report of the Commission of Inquiry into Legislation affecting the Utilisation of Manpower*, RP 32/1979, Government Printer, Pretoria.

South Africa (1980a) *Government Gazette No. 7246*, 3 October, Government Printer, Pretoria.

South Africa (1980b) *Population Census 1980. Selected statistical region Pretoria/ Wonderboom*, Report no. 02—80—22, Central Statistical Services, Pretoria.

South Africa (1983) *Report of the Technical Committee of Enquiry into the Group Areas Act, 1966, The Reservation of Separate Amenities Act, 1953, and Related Legislation*, Government Printer, Pretoria.

South Africa (1984a) *White Paper on the Promotion of Industrial Development as an element of co-ordinated Regional Development Strategy for Southern Africa*, Government Printer, Pretoria.

South Africa (1984b) *First Report of the Commission of Inquiry into Township Establishment and Related Matters*, RP/20, Government Printer, Pretoria.

South Africa (1985) *Report of the Committee for Constitutional Affairs of the President's Council on an Urbanisation Strategy for the Republic of South Africa*, P. C. 3/1985, Government Printer, Cape Town.

South Africa (1986) *White Paper on Urbanisation*, Government Printer, Cape Town.

South Africa (1987a) *Annual Report of the Department of Home Affairs 1985—1986*, R. P. 48/1987, Government Printer, Pretoria.

South Africa (1987b) *Report of the Committee for Constitutional Affairs of the President's Council on the Report of the Technical Committee, 1983 and Related Matters*, P. C. 4/ 1987, Government Printer, Cape Town.

South Africa (1988a) *Cape Metropolitan Area Guide Plan, Vol. 1, Peninsula*, Department of Development Planning, Pretoria.

South Africa (1988b) *Regional Development in Southern Africa and an exposition of the composition, aim and functions of the NDRAC, RDACs, RDAs, and DDA*, Supplement to *Informa*, Department of Development Planning, Pretoria, 2 April.

South African Institute of Race Relations (SAIRR) (1959) *Report by Research Officer on the Group Areas Proclamation for Kimberley*, Report no. 181/59, SAIRR, Johannesburg.

Stadler, A. W. (1979) Birds in the cornfield: African squatter movements in Johannesburg, 1944—1947, *Journal of Southern African Studies*, Vol. 6, no. 2, pp. 93—124.

Stadler, A. W. (1981), 'A long way to walk': bus boycotts in Alexandra, 1940—1945, in

P. Bonner (ed.) *Working Papers in Southern African Studies*, Ravan, Johannesburg.

Surplus People Project (1983) *Forced Removals in South Africa: The Surplus People Project Volume 5: The Transvaal*, Cape Town.

Sutcliffe, M. O. (1989) Housing. Appendix 10 in *The Durban Functional Region: planning for the 21st Century*, Report no. 1, Tongaat-Hulett Properties Ltd.

Sutcliffe, M. O. and McCarthy, J. J. (1989) Basis of estimation of the current population and projection to 2000. Paper prepared for *The Durban Functional Region: planning for the 21st century*, Report no. 1, Tongaat-Hulett Properties Ltd.

Sutcliffe, M. O., Todes, A. and Walker, N. (1989) Managing the Cities: An Examination of State Urban Policies since 1986. Paper given at the conference on Forced Removals and the Law in South Africa, Kramer Law School, University of Cape Town.

Swanson, M. (1977) The sanitary syndrome: bubonic plague and urban native policy in the Cape Colony 1900–1910, *Journal of African History*, Vol. 18, no. 3, pp. 387–410.

Swanson, M. W. (1983) 'The Asiatic Menace': creating segregation in Durban, 1870–1900, *International Journal of African Historical Studies*, Vol. 16, no. 3, pp. 401–17.

Swilling, M. (1990) The money or the matchbox, *Work in Progress*, Vol. 66, pp. 20–25.

Swindell, K. (1988) Agrarian change and peri-urban fringes in Tropical Africa, in D. Rimmer (ed.) *Rural Transformation in Tropical Africa*, Belhaven Press, London.

Tankard, K. P. T. (1985) *East London: The Creation and Development of a Frontier Community*. Unpublished M. A. thesis, Rhodes University, Grahamstown.

Tanser, G. H. (1965) *A Scantling of Time: the Story of Salisbury, Rhodesia 1900–1914*, Pioneer Head, Salisbury.

Tanser, G. H. (1974) *A Sequence of Time: the Story of Salisbury, Rhodesia 1900–1914*, Pioneer Head, Salisbury.

Technical Management Services (1988) The Supply and Demand of Coloured Housing in the Western Cape Regional Services Council area, City Planner's Department, Cape Town.

Teedon, P. and Drakakis-Smith, D. (1986) Urbanization and socialism in Zimbabwe: The case of low-cost urban housing, *Geoforum*, Vol. 17, no. 2, pp. 309–24.

Theart, C. (1990) *Towards a Guide Plan for East London/Kingwilliamstown*, Directorate of Planning and Engineering Services, East London.

Thomas, W. (1990) The Western Cape economy on the Way to 2000. Paper given at Development in the Western Cape Regional Congress, Cape Town, February.

Todes, A., Watson, V. and Wilkinson, P. (1987) Local government restructuring in South Africa: the case of the Western Cape, in R. Tomlinson and M. Addleson (eds.) op. cit.

Tomlinson, R. (1990) *Urbanization in Post-Apartheid South Africa*, Unwin Hyman, London.

Tomlinson, R. and Addleson, M. (1987) *Regional Restructuring under Apartheid: urban and regional policies in contemporary South Africa*, Ravan, Johannesburg.

Transvaal (1922) *Report of the Transvaal Local Government Commission*, TP 1/1922.

Truluck, T. and Cook, G. P. (1991) Changes in attitudes and actions in the Bokaap, Cape Town, *Contree: Journal for South African Urban and Regional History*, Vol. 26 (forthcoming).

Trump, M. (1979) The clearance of the Doornfontein yards and racial segregation, *Africa Perspective*, Vol. 12, pp. 40–56.

Turrell, R. V. (1984) Kimberley's model compounds, *Journal of African History*, Vol. 25, no. 1, pp. 59–75.

Turrell, R. V. (1987) *Capital and Labour on the Kimberley Diamond Fields, 1890–1971*, Cambridge University Press.

University of Natal (1952) *The Durban Housing Survey*, University of Natal Press, Durban.

Urban Foundation (1990a) *Policies for a New Urban Future: Tackling Group Areas Policy*, Urbanization Unit, The Urban Foundation, Johannesburg.

Urban Foundation (1990b) *Policies for a New Urban Future: Urban Debate 2010: 1 Population Trends*, Urban Foundation, Johannesburg.

Urban Foundation (1990c) *Policies for a New Urban Future: Urban Debate 2010: 2 Policy Overview: the Urban Challenge*, Urban Foundation, Johannesburg.

Utete, M. B. (1979) *The Road to Zimbabwe: the Political Economy of Settler Colonialism, National Liberation and Foreign Intervention*, University Press of America, Washington D.C.

Van Aswegen, H. J. (1968) *Die verhouding tussen blank en nie-blank in die Oranje-Vrystaat, 1854–1902*. Unpublished D. Phil. thesis, University of the Orange Free State, Bloemfontein.

Van der Merwe, I. J. (1985) 'n Ruimtelike en sosio-economiese vergelyking van werkritpatrone in Kaapstaad. Paper given at the Society for Geography Conference, Stellenbosch, July.

Van der Reis, A. P. (1983) *Black Commuting in Pretoria: Attitudes Towards Frequency, Punctuality and Waiting Times*. Special Report BCP 12, National Institute for Transport Planning and Road Research, CSIR, Pretoria.

Van Onselen, C. (1982) *New Babylon, New Ninevah: Studies in the Social History of the Witwatersrand*, Ravan, Johannesburg.

Van Niekerk, P. (1990a) Mixed fortunes, *Leadership*, Vol. 9, no. 1, pp. 37–44.

Van Niekerk, P. (1990b) Precious little, *Leadership*, Vol. 9, pp. 67–73.

Van Tonder, D. (1990) Gangs, councillors and the apartheid state: the Newclare squatters movement of 1952, *South African Historical Journal*, Vol. 22, no. 2, pp. 82–107.

Van Wyk, W. (1984) Personal communication (Bophuthatswana Department of Urban Affairs).

Voges, E. (1984) *Accessibility, Transport and the Spatial Structure of the South African City: an Historical Perspective*, National Institute for Transport and Road Research, Pretoria.

Von Garnier, C. (ed.) (1986) *Katutura Revisited 1986*, Roman Catholic Church, Windhoek.

Walt, E. (1982) *South Africa: A Land Divided*, Black Sash, Johannesburg.

Watts, H. L. and Agar-Hamilton, J. A. I. (1970) *Border Port: a study of East London, Cape Province, with Special Reference to the White Population*. Occasional Paper no. 3, Institute of Social and Economic Research, Rhodes University.

Wessels, J. (1989) Black housing in South Africa: the need for shelter. Confidential paper given at the Newick Park Conference, Sussex, 26–29 July.

Western, J. (1981) *Outcast Cape Town*, George Allen and Unwin, London.

Western, J. (1984) Autonomous and directed cultural change: South African urbanization, in J. Agnew, J. Mercer and D. Sopher (eds.) *The City in Cultural Context*, Allen and Unwin, London.

Western, J. (1985) Undoing the colonial city?, *Geographical Review*, Vol. 75, no. 3, pp. 335–57.

Western, J. (1986) South African cities: a social geography, *Journal for Geography*, Vol. 85, no. 6, pp. 249–55.

Wilkinson, P. (1983) A place to live: the resolution of the African housing crisis in Johannesburg, 1944–1954, in D. Hindson (ed.) *Working Papers in Southern African Studies*, Ravan, Johannesburg.

Williams, C. W. (1983) The theory of internal colonialism: an examination, in D. Drakakis-Smith and C. W. Williams (eds.), *Internal Colonialism: Essays Around a Theme*. Monograph 3, Developing Areas Research Group, Institute of British Geographers.

Wills, T. M. (1988a) The segregated city, in J. Laband and R. H. Haswell (eds.) op. cit.

Wills, T. M. (1988b) From rickshaws to minibus taxis, in J. Laband and R. H. Haswell (eds.) op. cit.

Wills, T. M. (1990) Segregation and desegregation: the impact of the repeal of the

Group Areas Act in Pietermaritzburg, *Proceedings of the Open City Symposium*, Pietermaritzburg 2000 Project, January.

Wills, T. M., Haswell, R. F., and Davies, D. H. (1987) *The Probable Consequences of the Repeal of the Group Areas Act for Pietermaritzburg*, Pietermaritzburg 2000 Project Report 16, City of Pietermaritzburg.

Wilson, F. (1975) The political implications for blacks of economic changes now taking place in South Africa, in L. Thompson and J. Butler (eds.) *Change in Contemporary South Africa*, University of California Press, Berkeley.

Winters, C. (1982) Urban morphogenesis in Francophone Africa, *Geographical Review*, Vol. 72, no. 2, 139–54.

Worger, W. H. (1987) *South Africa's City of Diamonds: Mine Workers and Monopoly Capitalism in Kimberley, 1867–1895*, Ad Donker, Johannesburg.

Younge, A. (1982) Housing policy and housing shortage in Cape Town 1942–1980, *Africa Perspective*, Vol. 21. pp. 9–28.

Zeitsman, H. L. (1980) Ruimtelike patrone van blanke residesiele mobiliteit in die Kaapstadse metropolitaanse gebied, *South African Geographer*, Vol. 8, no. 2, pp. 113–25.

Zille, H. (1989) Transforming the Apartheid City – The Potential in Cape Town. Paper given at University of Cape Town Summer School, January.

Zimbabwe (1984) *Annual Review of Manpower*, Ministry of Labour, Manpower, Planning and Social Welfare, Government Printer, Harare.

Zinyama, L. M. (1989) Multinational disinvestment: localization or socialist transformation in Zimbabwe's manufacturing sector?, *Area*, Vol. 21, no. 3, pp. 229–35.

ACTS OF PARLIAMENT

INDEX